# Multi-Agent Systems - Modeling, Control, Programming, Simulations and Applications

*Editor*

**Nicolas Radley**

scitus academics

**Multi-Agent Systems - Modeling, Control, Programming, Simulations and Applications**
Edited by **Nicolas Radley**

ISBN: 978-1-68117-313-9
Library of Congress Control Number: 2016937443


www.scitusacademics.com
Box No. 4766, 616 Corporate Way,
Suite 2, Valley Cottage,
NY 10989



Printed in United States of America on Acid Free Paper

# Preface

Multi-Agent Systems consist of multiple independent entities with different information and interests. The study of multi-agent systems (MAS) focuses on systems that interact with many intelligent agents. The agents are regarded as independent entities, such as software programs or robots. Their interactions can be mutually exclusive or cooperative. That means that agents can share a common objective (e.g. an ant colony), or their own interests can be pursued.

Multi-agent systems can be used to solve problems that for a single agent or a monolithic system are difficult or impossible to solve. Intelligence can include a methodical, functional, procedural approach, algorithmic search or learning to strengthen. Although the overlap is significant, a multi-agent system is not always the same as an agent-based model (ABM). The aim of an ABM is to find explanatory insights into the collective behavior of complying with simple rules, usually in natural systems, rather than solving specific practical or technical problems. Themes in which research on multi-agency systems can provide an appropriate approach include online trading, disaster response and social structures modeling. Research on multi-agent systems typically refers to software agents. However, robots, humans or human teams could also be the agents in a multi-agent system. A multifunctional system may contain combined teams of human agents. Agent systems are open and extensible systems which enable autonomous and proactive software components to be deployed. Multi-agent systems were developed and used in various fields of application.

This book, Multi-Agent Systems - Modeling, Control, Programming, Simulations and Applications, focuses on multi-agent technology, products and industrial applications.

# Table of Contents

# 1

# CHAPTER

# MULTI-AGENT SYSTEM FOR REAL TIME PLANNING USING COLLABORATIVE AGENTS

**Ana Lilia Laureano-Cruces[1*], Tzitziki Ramírez-González[2], Lourdes Sánchez-Guerrero[1], Javier Ramírez-Rodríguez[1]**

[1]*Departamento de Sistemas, Universidad Autónoma Metropolitana—Azcapotzalco, México D.F., México*

[2]*Maestría en Ciencias de la Computación, Universidad Autónoma Metropolitana—Azcapotzalco, México D.F., México*

## ABSTRACT

Autonomous agents are an important area of research in the sense that they are proactive, and include: goal-directed and communication capabilities. Furthermore each goals of the agent are constantly changing in a dynamic environment. Part of the challenge is to automate the process corresponding to each agent in order that they find their own objectives. Agents do not have to work individually, but can work with others and develop a coordinated group of actions. These agents are highly appreciated, when real time problems are involved, meaning that an agent must be able to react within a specific time interval, considering external events. Our work focuses on the design of a multi-agent architecture consisting of autonomous agents capable of acting through a goal-directed with: a) constraints, b) real-time, and c) with incomplete knowledge of the environment. This paper shows a model of collaborative agents architecture that share a common knowledge source, allowing knowledge of the environment; where we analyze it and its changes, choosing the most promising way for achieving the goals of the agent, in order to keep the whole system working, even if a fault occurs.

**Keywords:** Multi-Agent Systems, Blackboard Architecture, Planning, Schedule, Collaborative Systems, Cognitive Systems

# 1. INTRODUCTION

Each time we generate an artificial intelligent model, we are inspired by cognitive sciences. These allow us to understand how humans emulate the behavior of cognitive systems [1] [2]. Cognitive systems include both natural and artificial information processing systems responsible for perception, learning, reasoning, communication, action and more. One of the goals is to understand how natural and artificial processing devices could solve the most complex systems [1] [2]. In the case of the brain the robustness is completed by an ultra massive parallel processing in neural systems that can be compared to continue developing of very large scale integration. In order to get such complex computer systems it is necessary to think in the integration of these issues: (1) dynamics: how simple elements provide complex behavior over time; (2) self-organization: how an ultramassive parallel processing system organize themselves in a particular task without external intervention; (3) synchronization: how many processors operate without a signal (clock); (4) processing speed: the brain is fast even with slow processors; (5) power use: the power used by the brain to many processors; (6) timing: related to the ability of the brain to ensure everything happen on time; (7) robustness: the brain can tolerate fails in their components and arrive to his goal; (8) information representation in transmission: the brain communicates information between components in different ways; (9) construction and micro circuitry: the brain is organized in a complex hierarchy. This work is an approximation to the robust problem of simulate a human behavior, and is related with the (1), (2), (3), (6), and (7) issues that try to resolve a complex communication between processes. To achieve this, it is proposed to use reactive agents and their ability to build multi-agent systems. The agents cover a variety of topics including: a) distributed problem solving: the agent concept can be used to simplify the solution of large problems by distributing them to a number of collaborating problem-solving units; b) multi-agent systems (MAS): this research deals with appropriate ways of organizing agents. These include general organizational concepts, the distribution of management tasks, dynamic organizational changes like team formation and underlying communication mechanisms; c) autonomous agents: research on autonomous agents is primarily concerned with the realization of a single agent. This includes topics like sensing, models of emotion, motivation, personality, action selection, and planning;

d) a reactive behavior: an agent that follows another or reacts to his movements is perceived as an object that is driven by internal intentions and objectives. Based on the above idea, we define an agent as a computer system situated in some environment and capable of flexible autonomous action in this environment in order to achieve the objectives for which it was designed [3] . There are three questions related to the design of agents: 1) How can an agent determine the direction of your goals? The answer is our cognitive model that allows a view of the work as a whole, which enables us to view a model of collaborative agents. Each one independently carries out its work [1] . The abstraction in analysis and design is an important tool. Last make possible an appropriate level of description of the behavior, 2) How can an agent plan a series of actions to achieve these objectives and address? The interaction among agents are not pre-defined, there are not a global system goal. The agent has his own goals and capabilities. So in MAS the agents need to coordinate their activities and cooperate with each other. In the study case the planning is articulated in the design of behavior. Last is possible taking into account alternative behaviors that will be immersed inside the design of mental model. This is achieved by providing dynamic behaviors based in the stimuli presented in the dynamic environment. In this case using a shared memory, denominated blackboard, 3) How can an agent interact with other human and software agents to achieve their objectives? In the theory of MAS research there is a sub-line that refers to the different forms of communication that can be used in order to carry out a particular task. One way is through the changes that occur in the environment [4] . Each agent has its motivations and goals; each of them has different levels of competition. When presented in the dynamic environment stimuli that make a match with their responsibilities, the first make that trigger actions associated with different competences. Applications can be many: military, commerce, education, industrial processes, health, and expert systems in different areas. With this in mind this work proposes a fault tolerant MAS. This paper shows a model of collaborative agents architecture that share a common blackboard allowing a common knowledge source of the environment and the communication between agents through it, with soft kind restrictions. In this case, the system has available a planning, based on the earliest deadline first algorithm. If external factors (such as local faults) modify the behavior of the system during communication stage, it is important that the system keeps working successfully. In this proposal, the network control system is able to reconfigure itself to assure that the blackboard, which reflex the environment and has an own knowledge source, will be updated by messages between agents, and the task will be transmitted through some assigned communication channel using multi-agent approach. Each module in charge of this reassignment is represented by a multi-agent system that has the necessary knowledge to react to those eventualities, reassigning its transmission frequency in order to fulfill the service.

## 2. MULTI AGENT SYSTEM

In artificial intelligence, one of the hardest problems is to consider the real-world events, as input to the entity in charge of solving a particular problem. The latter is one of the reasons for the introduction of the agents in the field, where the real world with its dynamic changes, should be considered by the entity responsible for problem solving. This work introduces an algorithm based on artificial intelligence techniques like blackboard architecture, and collaborative and reactive agents. This permits distribute the decision making process into reactive autonomous agents all of them are incorporated in a MAS. This translates into time and the more time is available for the agent's computation, the more intelligent the behavior will become. Furthermore, the iterative improvement enables the planning process to easily adapt the plan to change or unexpected situation. In the case of the reactive agents, they do not have internal models of the environment and based its action in a cycle of stimulus-reaction, according to the events presented in the current state of the environment in which are embedded; hence, it could not have objectives, nor explicit planning mechanisms and still be able to solve problems classified as complex. However, this situation can be partially resolved by the different ways in which you can achieve the interaction between them and between them and the environment. In this work an agent is a real or abstract entity that is: 1) able to choose about himself and about the environment; 2) has a partial representation of the environment and 3) within a universe multi-agent is able to communicate with other agents. The behavior of this agent is the product of their comments, their objectives and their interactions with other agents. This form of approximation is based on the interaction and the purpose of making many specialists agents cooperate, instead of using a central program and general to solve a problem as it does the traditional artificial intelligence. System components emerge with a semantic skill obtained from a decomposition of the problem. This decomposition can meet the agents, their responsibilities and interactions. The skills acquired can be viewed conceptually as the points where decisions, such as levels of abstraction of a plan that is the solution of a problem or as partial solutions to solve the problem are taken. It follows that planning is in the analysis and design, hence it is so important. The internal system modeling is done based on the behavior you want to display the system. That is, we must have a clear and complete picture of what we want to do with the system. Otherwise it would be impossible to set sub-goals according to their tasks to each of the agents. This means that system designers with reactive agents must invest time in design to let precompiled or hard-wire fixed in the action-selection mechanism. And so it is concluded that while planning mechanisms (traditional artificial intelligence) leave much work to deliberative agent. Reactive deliberative approach leaves much work to the designer. According to [3] , in the analysis and design of a MAS we must take into consideration next aspects: 1) agent modelling: an agent need a model of its world; 2) multi-agent planning: in some cases agents will share plans in order to coordinate their behavior or to achieve a goal using others; 3) social relationships: agents may have social relationships, 4) interaction: agents may interact. In a multi-agent world in which interaction is not pre-defined, agents may need models of each other to decide how interact; 5) communication: agents may communicate to exploit interaction and ensure coordination. Taking into consideration last aspects next section are related with the analysis and design of the agent.

## 2.1. Agents

According to Smart framework proposed by [3] and resumed in Figures. First of all, we need to define a goal, which is a group of objectives to be fulfilled by an entity [3] . The agents must select their actions depending on the environment and their own and co-share goals with other agents. A goal describes a set of states, plus the attitude that the agent should have toward them. It is represented by a state made up of a set of affairs that can be carried out within the environment. Now we will define the agents within the aforementioned hierarchy (Figure 1). Here an agent is an instantiation object, which is associated with one or several goals. We have to consider that an agent goal can generate a passive or active reaction. Passive goals are goals which are imposed or assigned and active goal is how handles the actions to satisfy its goals interacting with the environment. An agent immersed in an environment has a set of perceptions producing the only attributes that the agents can get from the outside world; these perceptions depend on the agent capabilities and also on the world state. It is important to mention that an agent cannot perceive all the environment attributes; therefore, its behavior is based only on a subset of those attributes. This subject is related to the partial world representation proposed by Brooks [5].

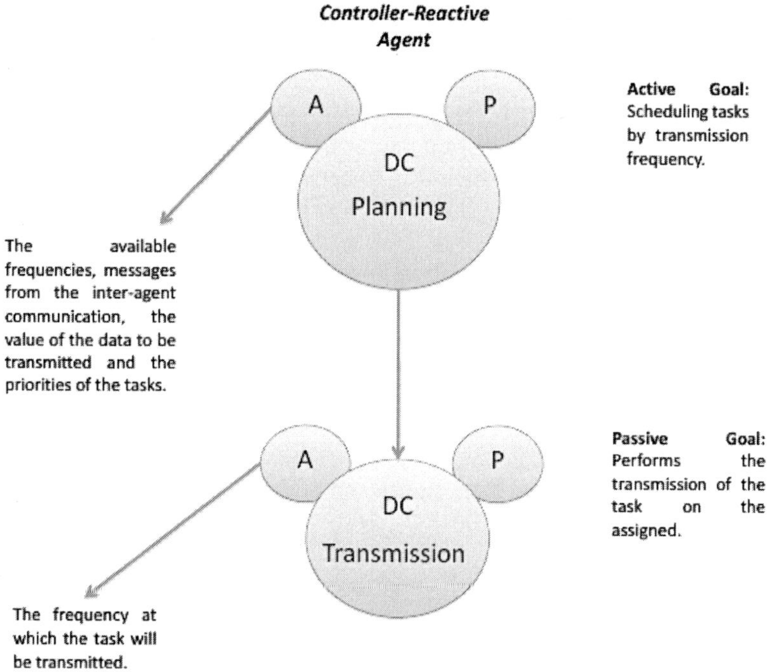


**Figure 1**. Controller Reactive Multi-Agent System.

## 2.2. Casc Study

According to the number of nodes in the system, it means that one MAS, which will call controller-reactive agent from now on; is assigned to each node in the network. The case study have two agents one active and the other passive as shown in Figure 1, it do not have presentation part, because all the proceeding is inside the software system.

Another important characteristic is the collaboration between the agents to select the most convenient option supported by the blackboard, due to this collaboration and the information of the blackboard architecture, the controller-reactive agent has a more real perception of the environment and is capable of a self-reconfiguration choosing its transmitting channel. In this way it ensures the transmission, getting a new frequency every time it needs to transmit. A real time distributed control system should execute its tasks at a specific time; it uses an algorithm to organize them named the

Planning [6] -[8] , it could be, for instance, earliest deadline first (EDF) to guarantee a good operation of the system. However there is no guarantee of this behavior; there are external factors that modify the operation at the communication stage [5] [9] [10] , so one way to guarantee that the information arrives to its destination is being aware of this variation. The goal of this work is to provide a way to have the information transmitted at different communication channels (according to several frequency transmissions); where, there is not a static frequency (it referrers to quantity of information transmitted in a unit of time) to transmit, instead of an option is selected from a blackboard, taking into account the restrictions and the agreement amongst agents. The following strategy is considered to overcome undesirable effects when collisions are present during communication between processes in a distributed environment. If a process needs to execute a set of tasks, which must be planned, one has to take into account that the sum of the times that each of them consume must be less than the specified time to run the process. If the tasks comply with this condition are said to be feasible to be planned. In this case it's a bandwidth planning, a task is likely to be planned if there is a suitable frequency within its bandwidth to ensure the time period in which it should be transmitted (to guarantee the real-time). The variables considered in each planner are: 1) the number of tasks; 2) the number of frequencies; 3) allowable range of frequencies for each task; 4) priority tasks (all tasks have different priority); 5) duration.

In a test schema where you have two agents with three tasks for planning each one, you can distinguish between two scenarios, in the first one all the tasks of two agents can be scheduled, in the second one there can be tasks that can be scheduled in this moment therefore the algorithm identifies them and does not allow that the information to be transmitted, as there is no available frequency, therefore they continue by selecting the best option for the other tasks. On having continued with the review of tasks that must be planned, it checks again if it can be scheduled in the current environment and continues with the procedure. Dynamic bandwidth scheduling allows the transmission of a task to a different speed each time, that is to say, the transmission frequency is changed using a parameter that indicates the period of time in which it is used. To notify to other agents the frequency that has been selected a message is sent and there are updated the blackboards of each of them. Our work is focused on performing the ordering of the communication between processes in a dynamic way, so the information can be transmitted on different communication channels to size the bandwidth and improve the performance of the system. That is, we use a static or predefined scheduling of communication, and allow each process to communicate through different channels, taking into account several restrictions, like: bandwidth, availability of frequencies and priority. In this paper distributed artificial intelligence is pursued based on collaborating multi-agent systems having a shared memory, by simplicity we call it, blackboard (black board architecture) [11] [12]. To achieve this goal, collaborating reactive agents are used at the scheduling stage in a distributed control system in order to implement an agreement with respect to the use of several communication channels (see Figure 2). Each agent performs a local scheduling algorithm supported by the blackboard information, which is modified periodically reflecting the state of the environment in which they are acting. If a controller-reactive agent needs to modify the messages sent through a specific communication channel, this is agreed with the rest. If this is a valid option the related agent is capable of schedule its transmission. This is implemented by the use of a blackboard among reactive agents in order to get a consensus with respect to the use of the available channels on the bandwidth.

Each node is composed by a controller-reactive agent, which performs the scheduling of tasks, assigning an adequate frequency for each process in order to transmit. To determine if the related communication channel is the adequate, the controller-reactive agent produces a selection based on the information at the blackboard, [6] [11] [13] [14]. The selection mechanism consists of a consultation among controller-reactive agents to know which communication channel is available, according to the interval assigned per process and the related global scenario. For the case of collision there is a mechanism based on the priorities of the tasks and the identifier of each controller-reactive agent. On the other hand we must remember that we have more channels than agents. The latter ensures that the message will always be transmitted. It is important to mention that competition focuses on the fastest channel. The blackboard is a table that reflects the state of the environment. In order to assign a communication channel, each controller-reactive agent reviews the general information contained in the blackboard and reach an agreement based upon two key issues: the flexibility of local processes to perform communication and the pursued scenario with these modifications. This blackboard contains the frequencies assigned to every controller-reactive agent per local process. So every certain period the controller-reactive agents are looking at and modifying this table if needed. In case of collisions, if the chosen channel is busy, the process is re-scheduled to choose another communication channel with respect to the available range assigned per process. The adaptation capacity represents a robust way of adaptability and predictability. The last idea drives the system to maintain or improve its operation. The final goal is to get a balance between the dynamic environment and a decentralized process in the system core [14] [15].

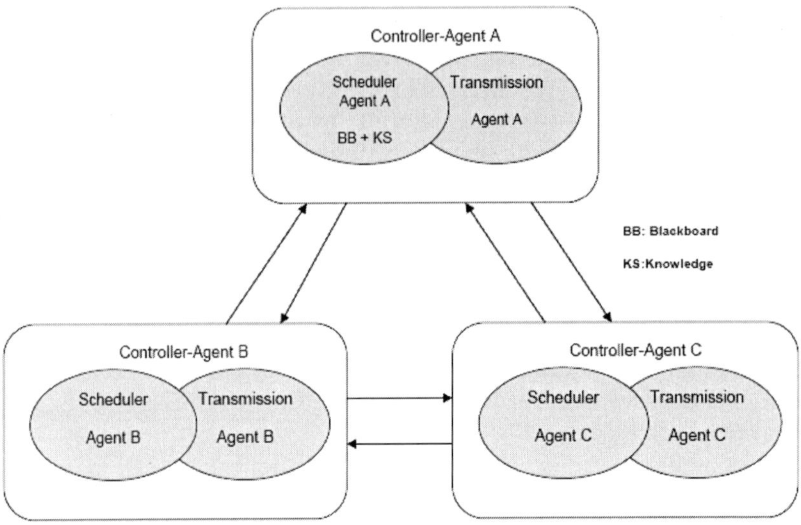


**Figure 2**. Scheduler inside of a Controller-Reactive Agent Anatomy.

## 2.3. Collaborative Systems

This paradigm refers to have independent modules that cooperate in order to solve a problem together. The details and complexity associated with each module is encapsulated within them. Problems are solved in a distributed way; this is accomplished by dividing the knowledge of the solution among the different modules [9] [16] [17]. Depending on the different ways of communication modules, we have: 1) strongly coupled systems, where communication takes place through shared memory; and 2) weakly coupled, where communication is through message passing. The case of strongly coupled systems is based on blackboard architectures. This kind of processing is driving by data and by the target, having at the end the problem performed. In the other side the case of weakly coupled systems, the communication among modules is achieved through message passing. There are different ways of message passing: 1) coherence based on cooperation, negotiation-based agreement; 2) hierarchy and organization; 3) computing and communication [4] [13] - [17]. In the case study we conceive a hybrid module that combines the two aspects of weakly coupled systems and strongly coupled systems. Corkill [16] proposes a very flexible way to combine the blackboard architecture with the multi-agent architecture. Taking into account the characteristics of collaborating processes. The case study is based on a Full-Fledged Blackboard Agents; where there are different reactive agents with its own knowledge source, and its own blackboard. For further information review the Corkill document. With this approach it is designed a model that allows autonomy based on the ability to share data. According to Corkill [16] , a collaborative system must have the following 6 features: 1) representation, having a sufficient number of modules that allow the understanding of the system as a whole; 2) incorporation, the modules are able to recognize the stimuli according to their skills; 3) emergency modules provide the ability to find the information they need to react in the environment; 4) interaction, create modules that are able to use the concurrent work of others while they are working in a shared task; 5) integration, providing them with a mechanism that allows them to combine and interpret the results obtained by other modules; 6) rationale, design modules capable of doing the right things at the right moment. This work has a model that allows take advantage of essential elements from next two approaches: 1) share data and; 2) have autonomy. These approaches make possible have modules with an emergent behavior. Taking into account the actions of other agents [11] [14] - [16] [18]. In order to design behaviors and achieve the six characteristics of a collaborative system, the controller-reactive agent is designed considering the detailed of the SMART Framework proposed by d'Inverno and Luck [3]. It provides us a classification based on objects, agents and autonomous agents [14] [16].

# 3. BASIC IMPLEMENTATION

We propose a collaborative system based on a shared memory and multi-agent architecture taking into account that communication need share a set of data; that allows triggering different actions in the different controller-reactive agents. These mechanisms generate the autonomy. According to the theory developed by d'Inverno and Luck [1] [3] [16], a multi-agent system is one that contains a collection of two or more agents. Since goals cannot exist without having been generated by autonomous agents, it is impossible for agents to exist without autonomy somewhere in the system. So the definition of a multi-agent system is any system that contains: 1) two or more agents; 2) at least one autonomous agent and 3) at least one relationship between two agents where one satisfies the goal of the other. For this case the multi-agent system has an autonomous one, the planning agent, and a passive one, the transmission agent (Figure 3). The blackboard is used by the autonomous agent in order to handle its autonomy, through the knowledge of the changes in the environment, which in the case under study refers to knowing which controller-reactive agent is using a given communication channel per process. For further clarification, see Figure 2 and Figure 3 [11] [12] [14] [16] [18] . A variant of the models proposed by Corkill [16] is used, where each autonomous agent (planning) has just one source of knowledge and a blackboard (see Figure 2). This MAS, called controller-reactive agent, allows: 1) communication, through the data written in each of the blackboards; 2) coordination, through the updating of such data; and 3) an autonomous behavior, due to the ability to perceive the dynamic environment through the data on the blackboard. In what follows we describe the distributed control problem and the proposed solution. In the proposed MAS, at each controller-reactive agent there is an autonomous agent that represents the knowledge source and a passive one that has the necessary capabilities to achieve the objectives of the autonomy [14] [15] [17] -[19] . Figure 3, shows the dependences between active and passive agents. Their functions are described using the notation of d'Inverno and Luck [3].

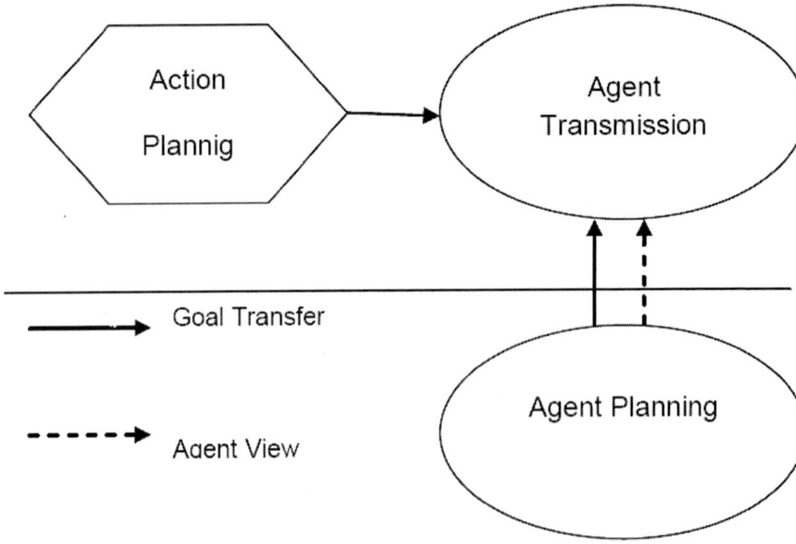


**Figure 3**. Dependence of agents in a Multi-Agent System.

**Planning: Autonomous Agent**

Type of Goal: Active

Objective: Scheduling tasks taking into account the transmission frequency.

**Perceptions**

Action it perceives: a) selects an adequate frequency for the transmission of a task, based on the record of available frequencies (blackboard).

b) Updates its record (blackboard) in accordance with the messages coming from other agents.

c) Modifies the frequency to transmit the task.

What can it perceive? The available frequencies, messages from the inter-agent communication, the value of the data to be transmitted and the priorities of the tasks.

What will it perceive? The adequate available frequency, in accordance to the priorities of the task at hand.

*Action*

Performs the planning of tasks to be transmitted in a dynamic way for a specific frequency, chosen from the revision of a shared memory (blackboard) that shows the possible options to transmit according to the allowed range, and updates itself checking the mailbox, were there is information about whether other agent is using or releasing a frequency range. In a collision case, the controller-reactive agent communicates with the other controller-reactive agent involved and defines the use of the frequency based on the priority of the task, updating the blackboard later on.

Type of Goal: Passive

Objective: Performs the transmission of the task on the assigned frequency.

Perceptions

Action it perceives: Transmits the task.

What can it perceive? The frequency at which the task will be transmitted.

What will it perceive? The frequency in which the task will be transmitted.

*Action*

Transmits the tasks in the frequency decided by the planning autonomous agent.

In this case, since we are dealing with frequency planning, we may say it will be able to be planned if there is an adequate frequency to warrant the time in which it must be transmitted (in real time) [6] [7] [8] [13] . The Planning agent may fulfill several actions depending on the variables that it is managing at an instant of time (t); 1) it verifies if there is a message in the mailbox. If there is a collision, i.e. if a chosen frequency is being used, a message is sent to the respective controller-reactive agent. Each controller-reactive agent knows the priority of the processes and the identifier of every controller-reactive agent. Taking into account the former data and the criterion based on the priorities of the processes, a channel is assigned and thus the assignment of frequencies is restructured from the Planning agent for this current scenario. In the agent's blackboard (Table 1and Table 2) there is a list of available channels that are related to the process willing to communicate and to the corresponding controller-reactive agent. The attributes of the Frequencies table (Table 1) are: frequency, task and agent. If the value of process is less than zero, it means that no communication channel has been assigned to the process; otherwise, there is a frequency assigned to the process (task).

Inside the same agent's blackboard there is a table of attributes per process, which shows the minimum and maximum frequency. The blackboard is updated as a result of the communication between agents [15] [16] [18] ; in this way each controller-reactive agent finds out the state of the others as well as its own. The attributes of the priorities table are: Process, Port, MinFrq and MaxFrq, Duration. For practical purposes it is assumed that the process is equal to the task priority, to avoid redundancy information. The Port column of the table shows if the process is a proper one (if the number is greater than zero) or the process of another controller-reactive agent (if it is less than zero) or is an available frequency (if it is equal to zero). The minimum and maximum frequencies, together with the priority number, allow us to select the frequency more adequate for each process, i.e., they will restrict and bind the range of frequencies in which the process can be transmitted. Duration shows the period of time on which the task will be transmitted in an assigned frequency.

# 4. THE MENTAL MODEL

In some areas it is possible to formulate cognitive theories of competition, specifying: what has to be calculated, when, and why, then on the basis of these theories an algorithm is developed to represent the mental model. This area of study is known as the theory of competition and is based on performing mental models. The essential part of mental models is its functional characteristic. They are defined as functional abstractions, with respect to the job or jobs, and provide a deductive framework for problem solving. The basic aspect of this doctrine lies in assuming that most of the inferences made daily depend on: spontaneous conclusions and a set of arguments that do not always lead to a valid conclusion. Once we have the design, the next step is to get the mental models; it is a pseudo-code that shows all the inputs, outputs and methods. The dynamic planning is based on a MAS supported by the Blackboard in a Full-Fledged scheme in order to share the information about environment. This configuration is due to it is required that agents must be independent in their self-performance, that means, they keep sharing information through the blackboard even if one of them stops working. Each agent, has it own Knowledge source (independent modules that contain the expertise needed to solve a problem) based on their competences and local views, that allows interpret the information according to their goals, executing the best action to achieve its mission. To execute a process we must take into account the sum of every period of time that are used to realize a task by each agent, which must be lower than the time specified for the process. If the every task fulfills these conditions, it means are feasible to be planned. In this case, as it is a bandwidth planning, a task is feasible to be planned if there is an appropriate frequency available to guarantee the task must be transmitted in time (Guaranteed Real Time). The Planning Agent must perform many actions: 1) if it is a task, verify if it has a message in the mailbox; if it is true, 2) selects a feasible frequency to transmit the task; 3) sends a message to the other agents notifying the selected frequency. If a frequency is schedulable, the frequency is changed (just before this, the mailbox is checked again to ensure that the frequency selected is available) and, at that time, that interaction between an autonomous and non-autonomous agents begins, the Transmission agent receives the task and the frequency, and executes the transmission returning the control to the Planning agent in the next instant of time.

**Table 1**. Frequencies (blackboard).

| Frequency | Task | Agent |
|---|---|---|
| 1 | −1 | −1 |
| 2 | 1 | 1 |
| 3 | 2 | 2 |
| 5 | 3 | 3 |
| 6 | −1 | −1 |
| 8 | −1 | −1 |
| 10 | −1 | −1 |

**Table 2**. Priorities (blackboard).

| Task | Min Frq | Max Frq | Duration |
|---|---|---|---|
| 1 | 2 | 10 | 10 |
| 2 | 2 | 10 | 20 |
| 3 | 2 | 10 | 30 |

**The planning agent mental model is:**
**Begin**
Initialize the Blackboard
Get the task from the queue tasks agent
Verify the system state
**If** (the agent must act due to the events generated in the environment)
Select the frequency from the Blackboard (Contemplates the Blackboard update)
**If** is a process feasible to be planned **then**
 **Begin**
  Verify the mailbox to guarantee the frequency is available
  Notify to the other agents that the frequency is being used.
  Modify the transmission frequency
 **End**
**Else**
  Get next task in queue (second step in the algorithm)
**End**

## 4.1. Planning Agent Attributes

The Blackboard has a list of available frequencies that are related to the task that is being transmitted.

The Frequency table attributes are: Frequency, Task and Agent. Task, is the task that the agent must transmit in a specific moment, here the duration is known and is the parameter that indicates when must be changed the transmission frequency. In the Blackboard (Table 1), if Task is less than zero indicates that is available, if not, indicates that the task is being transmitted. The Blackboard also contains a table for Frequency priorities (Table 2), which indicates the minimum and maximum frequency on the bandwidth, defining the priority for each Frequency of every task. Both, Task and Frequency are updated as a result of communication between agents, so, in this way the agents knows the status of others. A multi-agent architecture is based on the premise that there should be at least two autonomous agents and one other non-autonomous to help them to achieve their goals [5] . This proposal includes two agents, the first, Planning is autonomous and performs the bandwidth planning for each task and it assigns a specific frequency for transmission and manages the blackboard. It is the responsible of the communication with other agents by sending a message to the mailbox, updates the Blackboard if a message is received in the mailbox. Finally changes the frequency to transmit the task. The second agent is passive, and called Transmission, is who executes decisions from Planning agent and produces changes directly to the environment.

## 4.2. Planning Agent Methods

Planning selects and change the frequency transmission, sends messages to achieve communication between agents. All this is done through of five methods:

a) Select Frequency (SelectFrc). Analyze the Blackboard to get a feasible frequency to transmit a task. It depends on the priority and the minimum and maximum frequency on the allowed bandwidth. If a task is waiting for a frequency assignment, the mailbox is checked and the blackboard is updated, if it is a schedulable frequency, it sends a message to the other agents and executes ModifyFrc.

Inputs: Priority Task

Outputs: return the Frequency assigned or −1 if the task is not schedulable

b) Change Frequency (ModifyFrc); modifies the frequency where the task is to be transmitted.

Inputs: Task, frequency

Outputs: None

**The mental model to change the frequency is:**
If there are a task to be transmitted in a preselected frequency,
It is transmitted
It checks for another task in the queue

c) Send Message to the Mailbox (SendMail)

Sends messages to the mailbox in order to change information with other agents.

Inputs: Message

Outputs: None

The mental model to send a message is:

Sends the message to the mailbox

d) Check Mailbox (CheckMail)

Checks the mailbox in order to perceive the environment, interchanging messages with other agents.

Inputs: None

Outputs: Message, if there's no message return 0

**The mental model to check mailbox is:** gets message
If there are a message in the mailbox
returns message
if there aren´t message in the mailbox
returns empty mailbox

Transmission agent methods: transmission has only one method to transmit a task,

a) It Transmits a Task (Transmit)

Inputs: Task

Outputs: None

**The mental model to transmit a task is:**
It is enable output port
It transmits a task

# 5. EXPERIMENTS

The dynamic planning in a bandwidth, allows the task transmission in a different speed, see Figure 4. The frequency of transmission is changed using a duration parameter, which indicate the period of time, used in (see Table 2). Taking this into account, the parameters are defined and the Table 1 and Table 2 are built, while planning is carried out see Figure 4, it is noted as the transmission speed decreases and increases in run time, considering duration parameter. While the frequency is lower there are more pulses in the graph, it means the transmission speed is higher, in the other hand, while the frequency is higher, there are less pulses in the graph, it means the transmission speed is lower.

Figure 5, shows changes in frequency for each task, it is important to note that frequencies are not repeated at the same instant, it is possible due to the messages interchange between agents. In the first graph Task 1 is shown, where the frequency change is each second, the second graph shows Task 2 frequency changing every two seconds and the third graph shows Task 3 by the frequency change every three seconds. Whenever contemplate a new frequency selected frequencies being used for other tasks.

It should be noted, that Task 1 corresponds to agent 1, Task 2 to agent 2 an Task 3 to agent 3, and it is possible to plan n Tasks per agent.

# 6. CONCLUSION

The root of the problem comes from the distributed system in fault situations and its capability to be robust to these faults. In this case (distributed control system), special attention is placed on the communication aspect, and a dynamic planning is devised allowing it to organize the task communication, choosing a frequency at execution time. Due to the fact that the conditions of the system are dynamic, the assignment must be coupled to the environment. This proposal presents a strategy in which the information is transmitted in a flexible manner, since the value of the transmission frequency is no longer constant. There is a range of frequencies in which the process is allowed to transmit, making the management of frequency range options possible. It is worth mentioning that within the distributed system several local schedulers are required, which together perform the total planning of the system. The blackboard architecture allows displaying the partial solutions along the way to the global solution of the system; these solutions are those contributed by each one of the controller-reactive agents when making a choice for each process in which each one is taking part. At design time agents with blackboard included (Full-Fledged Blackboard Agents) was chosen [11], since it was required that collaborative multi-agent system be independent through the data they handle, allowing them to develop their controller-reactive agents, i.e., if one of them ceases to operate, this doesn't mean that others stop working, because in the case in point the system is distributed physically. With this in mind we model an hybrid module that combines the two main aspects of the two different multi-agent architectures. On the one hand a weakly coupled systems (messages passing) and on the other a strong-coupled system (blackboard) [11] [12] [16]. On the other hand, the multi-agent system warrants sending the information, tasks and messages for the communication among the controller-reactive agents. However, the one who warrants that the information gets to its destination is the communication protocol, which is being used [20] [21]. It is important to remark that this algorithm was thought for a real time distributed control reconfiguration systems, in order to manage all the faults that could appear in it, and then take one action to stabilize the system. In this case, if a controller-reactive agent stops its work, the others search for a free frequency and transmit its information in the selected frequency. That's the reason why this algorithm has a self-reconfiguration core. No matter if a fault occurs, the system get on working because its self-reconfiguration. Like we are using a multi-agent architecture we can add more parameters to be perceived from the environment (thought the autonomous controller-reactive agent), and more agents to execute specific actions [15] [17] [18] [22] . One use could be in the cognitive radio (CR) which is a wireless communication in which one element, the transmitter can detect which channels are busy and which are available to sends its information, through intelligent methods, all of this independently of the used technology [10]. According with Tachwali, Basma, Refai [23], other concepts important to take into account in the design of an effective cognitive radio are configurability and adaptability. Configurability is related with the ability to change the communication mode, as is inferred from the definition of cognitive radio. And adaptability enables an automatic communication mode selection based on wireless channel conditions. The demand for ubiquitous wireless services has increased during the last decade [3]; spectrum resources are limited and are allocated according to a fixed spectrum assignment policy. So another aspect to consider is the operation opportunistically in the frequency band allocated to a licensed system, on the primary user. In the case of cognitive systems [2] we make a contribution to a complex behavior basically because the design of the controller-reactive agents is based on a local view of different elements, last take into consideration the environment. The controller-reactive agents can have an organization without the external intervention. We propose a synchronize behavior based on hierarchy of the different processes. There is an assurance to achieve the final goal. Finally this algorithm is robust because it is tolerant to fails. The authors consider that the algorithm proposed is independent of the case study and could be implemented to: 1) improve the communication among process; or 2) to simulate human reasoning of an expert; given an additional positive aspect with reactive characteristics that implies autonomy.

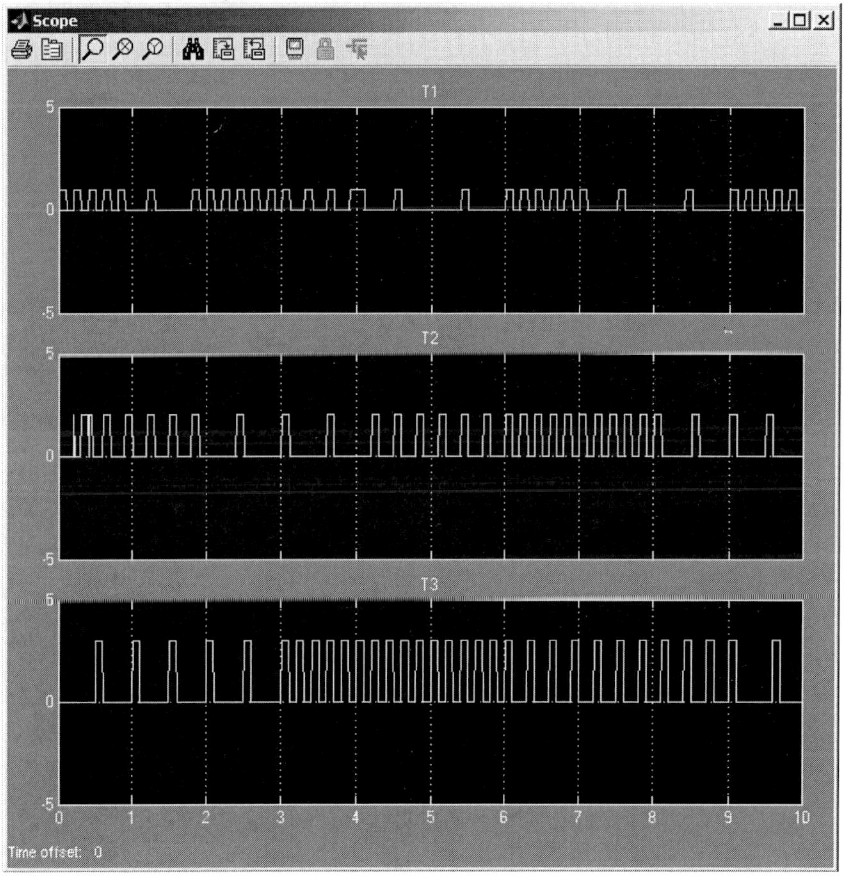


**Figure 4**. Frequency assignment.

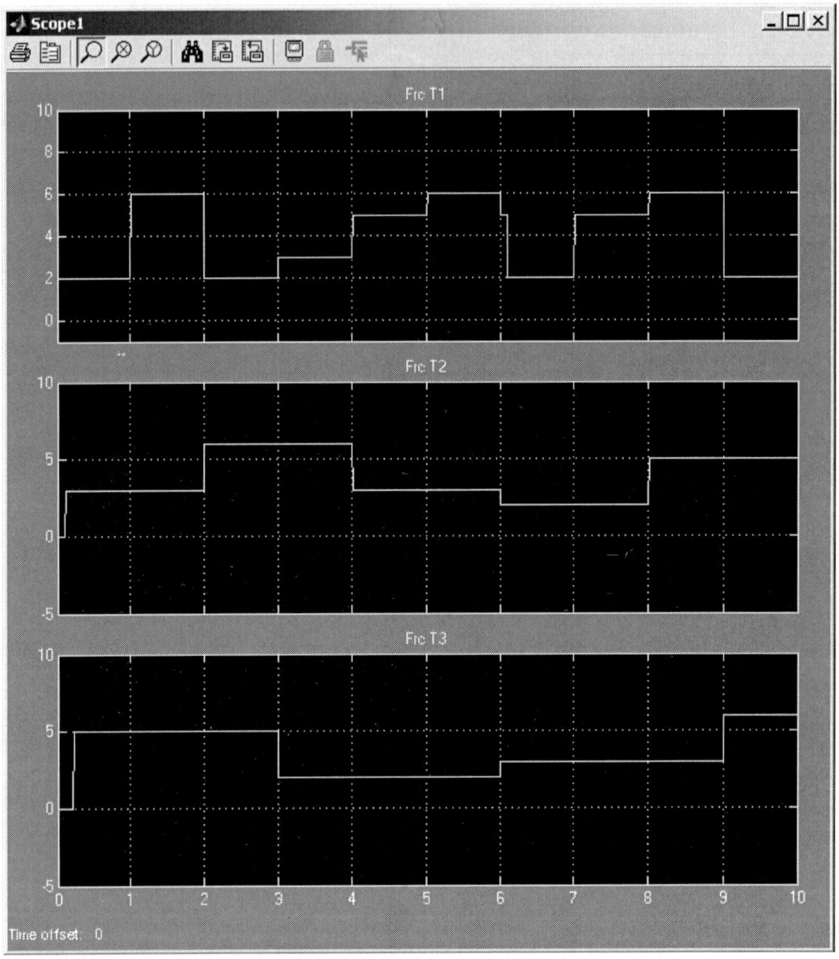


**Figure 5**. Time in the assignment frequency.

# ACKNOWLEDGMENTS

This paper is part of the research undertaken by Tzitziki Ramírez González to obtain the MSc. Degree in Computer Sciences at Universidad Autónoma Metroplitana-Azcapotzalco. Also as part of the Divisional project Soft Computing and Applications in the sub line Multi Agent Systems: Modeling and Simulation funded by the same University.

# REFERENCES


1. Laureano-Cruces, A.L., De Arriaga, F. and García-Alegre, M.C. (2001) Cognitive Task Analysis: A Proposal to Model Reactive Behaviors. Journal of Experimental & Theoretical Artificial Intelligence, 13, 227-239.

2. Quiñones-Reyes, P., Benitez-Pérez, H., Mendez-Monroy, E., Cardenas-Flores, F. and García-Nocetti, F. (2006) Reconfigurable Network Control using Fuzzy Logic for Magnetic Levitation Case Study. IFAC SAFEPROCESS.

3. d'Inverno, M. and Luck, M. (2006) Understanding Agent Systems. 2nd Edition, Springer Verlag, Berlin.

4. Laureano-Cruces, A. and Barceló-Aspeitia, A. (2003) Formal Verification of Multi-Agent Systems Behavior Emerging from Cognitive Task Analysis. Journal of Experimental & Theoretical Artificial Intelligence, 15,407-431.

5. Brooks, R. (1991) Intelligence without Representation. Artificial Intelligence, 47, 139-159.

6. Benítez-Perez, H. and García-Nocetti, F. (2003) Reconfigurable Distributed Control Using Smart Peripheral Elements. Control Engineering Practice, 11, 975-988.

7. Frey, D., Nimis, J., Wörn, H. and Lockermann, P. (2003) Benchmarking and Robust Multi-Agent-Based Production Planning and Control. Engineering Applications of artificial Intelligence, 16, 307-320.

8. Liu, J.W.S. (2000) Real Time Systems. Prentice Hall, Upper Saddle River.

9. Abdelzaher, T.F., Stankovic, J.A., Lu, C., Zhang, R. and Lu, Y. (2003) Feedback Performance Control in Software Services. Control Systems Magazine IEEE, 23, 74-90.

10. Bellahsene-Hatem, N.R., Mostefai, M. and El Kheir, A.O. (2011) Contribution au Diagnostic des évenéments défalliants dans un système Dinamique Hibride. Proceedings ICGST Conference on Computer Science and Engineering, Istanbul, 19-21 December 2011, 137-142.

11. Cervin, A., Henriksson, D., Lincoln, B., Eker, J. and Arzen, K. (2003) How Does Control Timing Affect Performance? Analysis and Simulation of Timing Using Jitterbug and True Time. IEEE Control Systems, 23, 16-30.

12. Ramírez-González, T., Quiñones-Reyes, P., Benítez-Pérez, H. and Laureano-Cruces, A. (2006) Reconfigurable Fuzzy Takagi-Sugeno Networked Control Using Cooperative Agents and xPC Target. Proceedings of International Symposium on Robotics and Automation (ISRA 2006), San Miguel Regla, Hgo, Mexico, 25-28August 2006.

13. Dong-Chan, O. and Yong-Hwan, L. (2010) Cooperative Spectrum Sensing with Imperfect Feedback Channel in the Cognitive Radio Systems. International Journal of Communication Systems, 23, 763-779.

14. Laureano-Cruces, A. and De Arriga-Gómez, F. (1998) Multi-Agent Architecture for Intelligent Tutoring Systems. Journal of Interactive Learning Environments, Swets & Zeitlinger, 6, 225-250.

15. Laureano-Cruces, A., De Arriaga, F., Ramírez-Rodríguez, J. and Escarela-Pérez, R. (2006) Control of Intelligent Learning Systems with Reactive Characteristics. Interactive Learning Environments, Swets & Zeitlinger, 14.

16. Corkill, D. (2003) Collaborating Software: Blackboard and Multi-Agent Systems & the Future. Proceedings of the International Lisp Conference, New York.

17. Jennings, N.R. and Bussmann, S. (2003) Agent Based Control Systems: Why Are They Suited to Engineering Complex Systems? IEEE Control Systems, 23, 61-73.

18. Laureano-Cruces, A. and Espinosa-Paredes, G. (2005) Behavioral Design to Model a Reactive Decision of an Expert in Geothermal Wells. International Journal of Approximate Reasoning, 39, 1-28.

19. García-Zavala, A. (2003) Propuesta de un método de planificación para la reconfiguración en línea de un sistema de tiempo real. Master Degree Thesis, Posgrado en Ciencia e Ingeniería de la Computación, UNAM.

20. Almeida, L., Pedreiras, P. and Fonseca, J.A.G. (2002) The FTT-CAN Protocol: Why and How. IEEE Transactions on Industrial Electronics, 49, 1189-1201.

21. Low, S.H., Paganini, F. and Doyle, J.C. (2002) Internet Congestion Control. IEEE Control Systems, 22, 28-43.

22. Quiñones-Reyes, P., Benitez-Pérez, H., Mendez-Monroy, E., Cardenas-Flores, F. and García-Nocetti, F. (2006) Reconfigurable Network Control Using Fuzzy Logic for Magnetic Levitation Case Study. IFAC SAFEPROCESS.

23. Tachwali, Y., Basma, F. and Refai, H. (2012) Adaptability and Configurability in Cognitive Radio Design on Small Form Factor Software Radio Platform. Wireless Personal Communications, 62, 1-29.

# 2

# CHAPTER

# A FRAMEWORK FOR QUALIFYING AND EVALUATING SMART GRIDS APPROACHES: FOCUS ON MULTI-AGENT TECHNOLOGIES

**Gillian Basso, Nicolas Gaud, Franck Gechter, Vincent Hilaire, Fabrice Lauri**

*IRTES-SET, UTBM, Belfort Cedex, France.*

## ABSTRACT

New needs and emerging societal constraints have put the emphasis on the inadequacy of the actual electrical grid. Indeed, it is impossible, or at least very hard, to 1) integrate renewable energy sources at a great scale within the actual electric grid, 2) enable communications between the various power suppliers and consumers, 3) design several different services that meet the needs of a wide range of end users. A key solution to these issues consists in using Smart Grids (SG). SG can efficiently control power flows by means of Information Technology (IT). Technically, a SG consists of a power system and a bidirectional communication system. Multi-Agent Systems (MAS) constitute a possible technology that can be applied to control and monitor the operation of power grids. Moreover, MAS exhibit distribution, adaptive and intelligent features. The goal of this paper is to propose a framework of qualification and evaluation for comparison SG approaches. First, a set of features of importance for smart grids definition is identified.

Then, in a second step, some criteria are given to evaluate the impact of SG on the society. Finally, these features are applied to existing MAS approaches addressing SG in order to understand and compare their different contributions.

**Keywords:** Smart Grids; Qualification and Evaluation; Multi-Agent Systems

# 1. INTRODUCTION

Nowadays, a paradigm shift occurs within energy systems [1]. New needs and constraints emerge, such as: reduced carbon dioxide emissions, greater introduction of renewable energy, diversification of power transactions, etc. These new needs and constraints have put the emphasis on the inadequacy of the actual electrical grid. Indeed, it is impossible, or at least very hard, to 1) integrate renewable energy sources at a great scale within the actual electrical grid, 2) allow communications between the various power suppliers and consumers, 3) design several different services that meet the needs of a wide range of end users.

A key solution to these issues consists in using Smart Grids (SG). SG can efficiently control power flows by means of Information Technology (IT), integrating not only the supply side but also the demand side and all the devices that allow the distribution of power. Technically, a SG consists of a power system and a bidirectional communication system. It mainly focuses on applying IT to the distribution and customer sides.

As defined in [2] smart grid is the term commonly used to refer to an electrical grid whose operation has been transformed from a twentieth-century analog technology based on the pervasive use of digital technology for communications, monitoring (e.g. sensing), computation, and control.

Emerging smart grid technology allows for finegrained sensing and control. This enables highly flexible, efficient, and optimized operation, including full support for market-based, demand-side management and the accommodation of alternative generation sources, including sources of consumer-generated electricity. A smart grid is defined as having the following seven principal characteristics, as specified by the US Department of Energy's National Energy Technology Laboratory in its modern grid strategy [3]. A smart grid:

- enables active consumer participation
- accommodates all generation and storage options
- enables new products, services, and markets
- provides power quality for the digital economy
- optimizes asset utilization and operates efficiently
- anticipates and responds to system disturbances
- operates resiliently against attack and natural disaster A similar smart grid vision is put forward in the document European Smart Grids Technology PlatformVision and Strategy for Europe's Electricity Networks of the Future [4].

Over the past ten years, many research projects have addressed the smart grids and the number of projects tackling smart grids increases every year.

The goal of this paper is to propose a framework of qualification and evaluation for a comparison of every approach for smart grids. The evaluation and comparison are based on the feature analysis approach [5]. First, a set of features of importance for smart grids is identified and defined. Then, in a second step, criteria is given to evaluate the impact of the grid on the society. Finally, these features are applied to existing MAS approaches addressing smart grids in order to understand and compare their different contributions. Among the heterogeneous approaches, some are based on the Multi-Agent Systems (MAS) paradigm. MAS are a good candidate to model and implement the intelligent components of a smart grid as they exhibit autonomy, reactivity, pro-activeness and collaborative capabilities [6,7]. A workshop dedicated to agent-based approaches for energy systems (ATES) has even been created as a forum of discussion for this research since 2010. That is why authors decide to compare only MAS approaches.

This paper is organized as follows: Section 2 presents an overview of an evaluation framework for smart grids, Section 3 details the qualification framework, Section 4 presents the evaluation framework and Section 5 provides a study of the main multi-agent approaches for smart grids.

## 2. EVALUATION FRAMEWORK OVERVIEW

Significant investments have been made, especially in European Union and United States, in order to develop, demonstrate and deploy smart grids technologies. Some recent works are trying to assess the costs, performance and benefits of smart grid technologies. There is currently no established standard in this area. Among the recent initiatives, we may notably mention:

- The 2nd EU-US workshop[1] on Smart Grid Assessment Methodologies jointly organized by the Joint Research Centre and the Department of Energy on the 7th of November 2011 in Washington DC.
- The Evaluation Framework for Smart Grid Deployment Plans [8].
- The smart grid scorecard[2] that provides a listing of the required features a product must have to be compliant with a smart grid.
- The incentives to determine reliable metrics, costs, and benefits from projects related to smart grids, like in [9] and [10].

The present article is part of this effort by providing an evaluation framework. The study focus on multi-agent approaches for smart grid. Though we decided to study multi-agent approaches for expertise reasons, it is worth noting that the proposed evaluation framework could also be used for approaches belonging to any other paradigms.

In the following two sections, we define a framework that decomposes the analysis and the comparison of smart grids into two steps. The first step aims at positioning the evaluated approaches and understand them. This step is divided into two dimensions:

Structural dimension represents the physical infrastructure of the smart grid. Family problem dimension defines the different classes of problems solved by the smart grid.

The second step, aims at evaluating quantitatively and qualitatively the different approaches by assessing the societal impacts of these approaches. The proposed criteria for evaluating the societal impacts are defined according to the main challenges of smart grids[3]:

Greenhouse gas reduction: How to reduce the carbon footprint of the overall supply chain? (addressed by Environmental dimension of the proposed framework)

Energy security: How to increase the network's capacity in order to manage a potentially diverse set of new requirements? How to provide interruption-free reliable power accommodating all generation and storage options, especially wind and solar power? (addressed by Structural and QoS dimensions)

Economic competitiveness & affordability: How to cost-effectively transition to a low-carbon energy system, increasing affordability? How to enable the new products, services, and markets through a flexible market to provide cost-benefit trade-offs to consumers and market participants? (addressed by Economic dimension)

Human integration: How to enable the consumers to actively participate by providing them with the choices and the incentives to modify their electricity purchasing patterns and behaviors (Smart metering, Advanced Metering Infrastructure)? (addressed by Human dimension)

For each of these challenges a set of indicators is proposed. It is not always possible to test each approach with every indicator or to obtain resources from the literature that provide results of such analysis.

# 3. QUALIFICATION FRAMEWORK DEFINITION

In the following, the different aspects of the framework are defined. These aspects are composed by the first (qualification) and second (evaluation) steps. The first step is detailed through two viewpoints or dimensions (structural and family problems) and the second step is detailed through the societal dimension. A list of criteria (written in bold face) is presented thereafter for each dimension (structural and family problem) with the corresponding definitions and explanations.

## 3.1. Structural Dimension—Industry Focus

The structural dimension defines the energy infrastructure of the smart grid. The energy infrastructure consists of two highly-interrelated and complex systems: the energy physical network and the energy market [11]. These two systems and their relationships are precisely described by the ontology in [12].

First, we briefly explain the main concepts of the electric power delivery system described in the ontology.

Second we highlight questions must be answered for completely describing a power grid.

- The producers are responsible for generating energy.
- The transmission system operators are responsible for operating the transport of energy on the high-voltage interconnected system, called the transmission network, in order the ensure long-term services, like the balance of supply and demand, for instance.
- The distributed system operators are responsible for operating the transport of energy on medium-voltage and low-voltage energy systems in given areas. They ensure the connections of the medium-voltage and low-voltage energy networks, called the distribution networks, to the high-voltage distribution network. They also ensure the energy delivery to customers, but without including supply.
- The suppliers are responsible for the sale of energy to customers. In a liberalized energy infrastructure, a supplier can be a wholesale customer who purchases a commodity (for instance energy) with the purpose to sell it subsequently.
- The households (or customers) purchase energy for their own use.

Physical Is the network for physical delivery taken into account?

Size What is the physical network size (network voltages, number of devices, power capacity)?

The size of a network is an intricate concept and cannot be defined precisely by a single value. For completeness, a network size must contain the different voltage of the network, and the number of devices present in the network.

Storages Which types of storage systems are used (total energy, power capacity, power variation)?

Sources Which types of sources are used (renewable, capacity factor, other characteristics defined by the ontology)?

Loads Which types of loads are used (predictable, disconnectable, other characteristics defined by the ontology)?

Scalability Is the physical network scalable in realtime?

Communication Is communication between devices possible?

The new energy infrastructures have to offer a communication between distributed appliances (for example, for a demand response [13]), that is they could allow to take into account priorities on load demands or could increase the balance between consumption and production.

Microgrids Is the physical network composed of microgrids?

Microgrids [14] can be defined as parts of the energy network that comprise intermittent sources.

This approach allows for local control of distributed generation, thereby reducing or eliminating the need for central dispatch.

Islanded Is the physical network islanded?

Network type Which type of physical network is used?

The proposed approaches are validated on the basis of experiments carried out on a simulated physical network, a real network or a combination of both.

Models If simulated, which models are used for the appliances?

Timescales If simulated, which timescales are taken into account in the energy network?

Managing energy in this system implies to take into account several strong constraints.

Besides, supply and demand occurring within the physical network must always be balanced on a short timescale (milliseconds or seconds).

Commodity trade Is the network for commodity trade taken into account?

Dynamic prices Are the prices allocated dynamically between actors?

Time frames Which timescales are taken into account in the energy market?

Depending on the possible control actions, the energy balance has to be maintained at the different time frame [15]. The wholesale market and the retail market ensure energy balance at medium time frame (minutes or hours) and long time frame (days, months or years). The balancing market ensures energy balance at small time frame (milliseconds or seconds) and medium timescale.

## 3.2. Family Problem Dimension—Domain Focus

As stated in [16] smart grids raise many significant challenges for the AI field. Indeed, concepts and techniques will be needed to solve the numerous problems that are not solved today by the current electricity grid. Among these problems, one can cite: the maintenance of stable voltages and frequencies, supply reliability, Distributed Energy Renewables (DER) management, heterogeneous and distributed actors, self-healing networks, ...

Follows some already well-identified problems.

Unit commitment The problem of unit commitment (UC) aims at finding the dispatch of available generation sources that meet the electrical load with the minimum cost. In other words, the aim is to determine the combination of available generating sources or units and to schedule their respective outputs to satisfy the forecasted demand with the minimum total production cost under the operating constraints enforced by the system [17].

Demand side management Energy demand management, also known as demand-side management (DSM) or Load Management [18], is the modification of consumer demand for energy through various methods such as financial incentives and education. Usually, the goal of demand-side management is to encourage the consumer to use less energy during peak hours, or to move the time of energy use to off-peak times such as nighttime and weekends. Peak demand management does not necessarily decrease total energy consumption, but could be expected to reduce the need for investments in networks and/or power plants by reducing demand peaks.

Demand Response The demand response (DR) is an extension to DSM problem. The difference lies in that demand response mechanisms respond to explicit requests to shut off, whereas DSM passively shut off when stress in the grid is sensed.

Supply and demand matching Supply and Demand Matching (SDM) is concerned with optimally using the possibilities of electricity production and consumption devices to alter their operation in order to increase the overall match between electricity production and consumption.

Vehicle to Grid The idea behind the Vehicle-to-Grid (V2G) concept, is to use the flow of power (both in input and output) of an electric vehicle. These flows can be real add-on for the electrical power grid. Indeed, these vehicles, either electric cars (BEVs) or plug-in hybrids (PHEVs), have the capability to produce AC electricity. The challenge raised by this idea is to provide electricity precisely when needed and recharge these vehicles efficiently. By communicating with the electrical power grid BEVs and PHEVs could then implement a Demand Response service.

Virtual Power Plant A Virtual Power Plant (VPP) is a cluster of distributed generation devices (such as microCHP, PV, wind-turbines, small hydro, etc.) which aggregate themselves to sell electricity. The goal of VPPs is to maximize the value for both the end user and the distribution utility by using software systems. They are dynamic, deliver value in real time, and can react quickly to changing customer load conditions. A VPP matches up a variety of distributed energy systems with intelligent demand response capabilities and aggregates those resources into an asset that acts like a centralized power plant. VPPs are similar to microgrids; however, while microgrids are very local in scope, VPPs can theoretically be deployed on a wide scale at the utility level with the ability to tap resources in real time, and with enough granularity, to control the load profiles of customers, aggregate these resources, and put them up on a trader's desk.

Self-Healing Network One of the major advantages of smart grids is to allow self-healing of the network without the intervention of technicians. This will ensure more reliable supply of electricity, and reduced vulnerability to natural disasters or attack.

# 4. EVALUATION FRAMEWORK

The points described in 3 present the upstream work for smart grid analysis. However, the objectives of the smart grids are to help companies and individuals to solve the problems due to the daily energy production and management. In order to evaluate the impact of the incoming smart grids, concepts must be defined, we integrate the concept of sustainability usually considered as a composition of Environment, Economy, and Society [19] and we also add a human dimension representing system's control and supervision capabilities.

We tried to collect them within an assessment framework combining these 4 different perspectives grouped under the umbrella term societal dimension.

## 4.1. The Environmental Approach

The most significant environmental impact of power generation is in the form of emission to air, ground and water. Of those emissions, the most significant in terms of impact are emission to air of carbon dioxide $(CO_2)$, nitrogen oxides $(NO_x)$, sulfur dioxide $(SO_x)$, methane $(CH_4)$, nitrous oxide $(N_2O)$ and particulates. The impact on water is more complicated, involving heat, volume and emissions. The objective of this part of our framework mainly consists in determining the benefits of a smart grid's model to reduce its carbon footprint. However, the objective can be extended to nuclear wastes or other discharges influencing environment. To reduce significant environmental impact, two main goals are easily focused:

- Reduce pollutants' emissions and wastes' production (reducing oil usage, reducing nuclear power plants, etc.).
- Lower transmission and distribution losses

## The Global Warming Potential (GWP)

It is therefore, important to have indicators to evaluate the performance of a smart grid in terms of environmental impact (e.g. [20, Chapter 11]). To estimate the greenhouse gases' emission impact, the United Nations Framework Convention on Climate Change[4] (UNFCCC) sets up a cost measure, based on the environment impact of the $CO_2$, the Global Warming Potential (GWP) [21]. With this cost measure, one can easily evaluate the greenhouse effect of every generator on the environment, given the amount of emitted gases during production, and thus it will be easier to compare them.

The difficulty of the usage of this cost measure remains in evaluating the amount of greenhouse gases' emission, but some research has been made in this problem, and it is currently possible to find the quantity of $CO_2$ emitted per kWh (as [22] or in [23]).

## 4.2. The Economic Criteria

As an economic aspect is already included in an energy network (see 3.1), it is important to define all costs, such as investments (CAPEX), ongoing costs (OPEX) but also the revenues implied by the investments to establish the interest of the problem to be resolved.

All costs and revenues are depending on the problem to be solved. As described in 3.1, the energy market contains 3 main kinds of actors with different goals. The producer side is the well-known part of the economic approach. The pertinent elements for this actor are the cost operation and the investments of energy generators. The distribution of the energy created by the new devices also implies a cost which has to be taken into account. Adding an intelligence into a grid implies potentially a communication infrastructure. It can be interesting to manage the energy loss during the transportation. For the consumer side, the main aspects are subscription, taxes and a price of the kWh. The first and the second can vary depending on the location of the consumers, while the third also varies during time.

In [24], the authors precisely describe costs and revenues of the electricity market.

The Asia Pacific Energy Research Centre [25] provides some indicators to assess of energy efficiency from an economic perspective. They propose economic value based energy-efficiency indicators measuring the quantity of energy consumed relative to the economic/onetary value of the activity generated, denominated in a currency-related unit. This is an extension of the study published by the French Agency for Environment and Energy Management (ADEME) [26], who suggests three alternatives for comprehensively reviewing trends in energy efficiency at a sectoral or sub-sectoral level:

### 1) Energy Intensity

Considers whether energy, as an input to production, is used efficiently. Energy intensity's analysis is generally based on relative comparisons with established benchmarks, historical trends, or other comparable energy intensities.

### 2) Techno-Economic Ratios

Calculate, from an engineering perspective, the economic production associated with the unit or specific consumption of energy.

### 3) Energy Savings Indicators

Endeavor to measure energy savings achieved by consumers over a period. These "techno-economic effects" essentially analyze changes in the techno-economic ratios.

In [27], authors provide a step-by-step assessment framework based on the work performed by the Electric Power Research Institute (EPRI) for conducting costbenefit analyzes of smart grid projects.

## 4.3. The Quality of Service Criteria

Nowadays, it would be impossible to imagine daily life without having a continuous access to energy. Offering this service to a growing number of persons and needs is very challenging. For instance, the transportation of the energy must meet numerous constraints in order to ensure power quality. Among these constraints, one can cite: nonzero frequency impedance, harmonic variations, etc.

Some statistical indicators of transmission and distributions circuits exist [28,29] for assessing the quality of such services, among the best known, we can mention:

**1) MAIFI**

Momentary Average Interruption Frequency Index measures the weighted average number of outages that last less than ten minutes, which occurred in a year and with reference to the total connected load.

**2) ASIDI**

Average System Interruption Duration Index measures the weighted average number of outages equal to or more than ten minutes, which occurred in a year and with reference to the total connected load.

**3) ASIFI**

Average System Interruption Frequency Index measures the weighted average number of outages equal to or more than ten minutes, which occurred in a year and with reference to the total connected load.

**4) SAIDI**

System Average Interruption Duration Index is the average outage duration for each customer served.

**5) SAIFI**

System Average Interruption Frequency Index is the average number of interruptions that a customer would experience.

**6) CAIDI**

Customer Average Interruption Duration Index (SAIDI divided by SAIFI).

**7) SISI**

System Interruption Severity Index measures the ratio of the unserved energy to the system peak load.

**8) FOT**

Frequency of Trippings per 100ckt-km measures the number of forced line outages (transient and permanent or sustain) initiated by automatic tripping of relay.

**9) FLC**

Frequency Limit Compliance refers to the percentage of time during the rating period that the system frequency is within the allowable range of 60 ± 0.3 Hz.

**10) VLC**

Voltage Limit Compliance refers to the percentage of the number of voltage measurements during the rating period that the voltage variance did not exceed ±5% of the nominal voltage.

Furthermore, some standards exist such as [30], which offers practice developed out of an increasing awareness of the difficulty in comparing results obtained by researchers. The Council of European Energy Regulators provides an extensive analysis on the quality of energy supply throughout Europe [31]. It provides a collection of indicators monitoring continuity of supply (interruptions), voltage quality and commercial quality used to develop a complete benchmark.

## 4.4. The Human Integration Criteria

From a consumer's point of view, power grid remains a black box, an unlimited amount of energy. Unfortunately, unlike fossil-fuel or nuclear based power generation, most renewable-energy sources depend upon generally unpredictable environmental factors (solar, wind, etc.). Thus, it does not suffice to replace coal or nuclear plants with solar power plants to ensure a reliable and stable energy production as consumers are now accustomed. The advent of smart grid will also require the change in consumer behavior.

This change in behavior especially requires new interface design and new ways to present information about energy usage to the user. Numerous studies have shown the impact of how to present information about energy usage impacts consumer behavior [32-37]. It appears that human behavior with respect to energy cannot be modeled only in terms of cost-effectiveness. It is directly influenced by the following factors [32]:

- Personalized information according to a user specific configuration.
- Goals and commitments.
- Constant feedback, particularly with measures of progress toward goals.
- Financial Incentives to encourage consumers to participate Supervisory Control And Data Acquisition (SCADA) are a first step towards a monitoring interface that could participate in user behavior change.

IEC PC 118[5] is working on a standard covering the architecture of smart grid user interface, function and performance requirements of demand side systems, information exchange interfaces among demand side systems/equipment, with the aim to support applications, such as a demand response.

## 4.5. Summary Statement

This evaluation Framework gives 4 criteria to evaluate the impact of smart grid on the society. In this section, we will describe numerical criteria to evaluate and compare approaches solving the same problem in the same structure. The criteria defined below are not the only to evaluate a grid but they represent the improvement of the future grid as defined in the section 2.

- The Global Warming Potential (GWP) appears to be the best existing factor to evaluate the environmental aspect of a smart grid.
- The economic factor is depending of the vision of the future grid, but all aspects rely on the price of the energy. The money (dollar, euro, etc.) is thus the obvious factor to compare several grids.
- The Quality of Service (QoS) has two perspectives, the first is the feeling of the end-users about the grid (with MAIFI, CAIDI, etc.) and the second

is the network stability (with FLC, VLC, etc.). These perspectives are linked, as the first can be seen as the result of the second.

- The Human Integration (HI) is the hardest criterion to quantify. The simplest way to evaluate the human interaction with the grid is the number of settings that users can modify.
- Smart grids of the same domain (see Section 3) can be evaluated quantitatively and qualitatively following these criteria to offer a comparison of the result.

# 5. MAS APPROACHES

In this section, we conduct a comparative study of the main multi-agent approaches for smart grids using the qualification and evaluation framework previously defined. After over ten years of research, indisputable progress has been made. Indeed, recent approaches are more and more efficient but the benchmark results show that many efforts are still required to offer a truly comprehensive toolbox of approaches for smart grids.

The qualification of the framework can globally be set even if the structural dimension is not completely defined. On the other side, the evaluation framework cannot be used with the current analysis of the studied technologies because of the lack of numerical data. This is one of the point highlighted by this paper.

## 5.1. GridAgent

Developed by Infotility, the Grid Agent [38,39] Enterprise Agent Manager (EAM) Suite is the main user interface which provides centralized management and works in conjunction with a suite of pre-configured agent types, specialized editors, applications, and reporting alerting tools, including: An out-of-the-box, complete agent management collection included specialized property editors: RateEditor, ModelEditor, and RuleEditor, PlanningEditor Preconfigured visual information and analytic tools include: EventManager, Smart Dashboard, Resource Viewer, SmartFilter, Cost Manager.

The GridAgent framework implements several types of agents: "Sense and Control" and "Resource" Agents who have analytic methods to calculate optimal response based on pricing signals. "Planning and Optimization Agents manage Distributed Energy Resources (DER) devices under various operational scenarios such as optimal microgrid control strategies. "Blackboard" Agents can store databases (from several media, like Internet or the MAS).

GridAgent also offers human interactions and network protection.

This Suite is developed for managing distributed energy resources and can be used for large-scale integration of distributed energy and renewable-energy resources into real distribution systems.

Its evaluation following the given criteria are in table 1.

**Table 1**. Grid agent evaluation.

| | **Structural Dimension** |
|---|---|
| Physical | Yes |
| Size | The system is created to manage energy network supplying 24,000 network transformers in 63 secondary networks. It has to manage the 3G/SOF including the DER, consider substation load transfer, application of HVDC, higher distribution voltage (27 kV, 33 kV, 33+kV), minimal secondary voltage, company-deployed mobile DER assets |
| Storage | smart charging of PHEV |
| Sources | Dispatch of distributed and renewable generation |
| Loads | Loads present in buildings like home or company (e.g. HVAC) |
| Scalability | Yes, agents are "plug and play" |
| Communication | Yes, agents play roles, which are the ability to send and receive set of messages, FIPA compliance, could use IP |
| Microgrids | Yes, the system enables the ability to create on-the-fly "aggregate" blocks of capacity for presentation to the energy markets |
| Islanded | Yes, the system enables the ability to create intelligent islanded grid |
| Network type | Real, an example is presented in [39] |
| Commodity trade | Yes |
| Dynamic prices | Yes, the demand response algorithm is based on market transactions |
| Time frames | unknown |
| | **Family Problem Dimension** |
| Demand Re-sponse | Aggregation of capacity (*i.e.* consumption). Ability to shift or curtail load or dispatch local generation. Assume that each agent has a planning execution cycle |
| | **Societal Evaluation Criteria** |
| Environment | Reduce energy consumption during peak thanks to load shifting and load shedding and also use Renewables. |
| Economic | Each agent in the system attempts to choose among available alternatives according to the criterion of minimum operating cost |
| QoS | Using DER, which has the potential to provide reactive power (generated close to loads) to maintain grid voltage stability |
| Human | expected to operate over a large range of hardware platforms and sites (such as PDA) |

## 5.2. Homebots

Homebots [40,41] is an approach that deals with the smart distributed equipment management in a house. It is based on a multi-agent system, the agents of which are directly linked to one specific equipment. The management process is based on a computational market, which can be considered as a multi-agent systems sub-field, involving specific agents.

In this context, every load is represented by an interactive load agent, the preference and needs of which are translated into a utility function. The agents are grouped taking into account the topology of both the electrical grid and the communication system.

The utility function is embodied by a utility interface agent. This helps to provide an interface between the utility and the load management system even if the time scales for the different kind of agents (load and utility) are not necessarily the same. In the market management paradigm, the utility interface agents can be considered as a local utility salesperson. This utility agent can thus be directly manipulated in order to make the system go in on a specific way. The utility function is computed taking into account several elements such as: load model, load's current state, utility contract, user model, expected evolution of the local market.

The Homebots evaluation following the defined criteria is made in table 2.

## 5.3. IDAPS

IDAPS [42] is a distributed smart grid concept proposed by Advanced Research Institute of Virginia Tech. The agents in the IDAPS MAS work in collaboration to detect upstream outages and react accordingly to allow the microgrid to operate autonomously in an islanded mode. The proposed MAS consists of:

- A control agent who monitors the system voltage, detects problems and sends signals to the main circuit.
- A DER agent who is responsible for storing associated DER information and monitoring and controlling DER power levels.
- A user agent who acts as a customer (user or load) gateway.
- A database agent who is responsible for storing system information.

**Table 2**. Homebots evaluation.

| | **Structural Dimension** |
|---|---|
| Physical | No |
| Size | From one house to several houses. |
| Storage | Not defined. |
| Sources | Not defined, the only source for the system is the main grid. |
| Loads | House classical loads (wash machine, lights, etc.). Each is represented by one agent who decides to connect the load or not. |
| Scalability | yes, depending on the loads used. |
| Communication | The communication is made indirectly through the market exchanges. |
| Microgrids | Indirectly yes but not really mentioned |
| Islanded | No |
| Network type | In [41], both are mentioned but only simulations are presented. |
| Models | everything is integrated into utility function. |
| Timescales | Not detailed (expressed in time units) |
| Commodity trade | Yes |
| Dynamic prices | Yes. Homebots is based on market computational market economies. |
| Time frames | The utility function includes both local agent timescale and market timescale in the prediction that has to be taken into account. |
| | **Family Problem Dimension** |
| Adaptation to power production | Two examples of adaptation to power shortage are presented. In the first, the adaptation is made thanks to utility function. In the second, the adaptation is made by influencing the energy price. |
| | **Societal Evaluation Criteria** |
| Environment | Not available. |
| Economic | Minimizing energy cost based on market evolutions. |
| QoS | Not taken into account |
| Human | Human interaction can be made indirectly by influencing the market. |

IDAPS is realized with Zeus [43] multi-agent system platform, which is FIPA-compliant.

This work aims at demonstrating a practical implementation of multi-agent systems in a smart grid located at a distribution level. It also demonstrates that the agent's capability can be considered as a software alternative to a traditional hardware-based zonal protection system for isolating a microgrid. IDAPS separates the multi-agent system (developed with Zeus) and the microgrid hardware (developed with Matlab/Simulink).

The evaluation of IDAPS is presented in table 3.

## 5.4. IDEaS Project

The IDEaS Project has been focusing on the following problem families since its advent:

- Adaptive Home Heating ([45])
- Demand Prediction ([46,47])
- Demand-Side Management ([48,49])
- Electric Vehicles ([50-52])
- Energy Exchange ([53-55])
- Exposure in Auction Markets ([56])
- Micro-storage ([57,58])
- Virtual Power Plants ([59])

**Table 3**. IDAPS evaluation.

| | **Structural Dimension** |
|---|---|
| Physical | Yes |
| Size | Circuit 9 of Virginia Tech: a residential distribution circuit at or below 12.47 kV, 34 laterals with 117 transformers serving 780 customers. |
| Storage | Battery Model from Matlab-Simulink 2009 [44] |
| Sources | Solar photovoltaic, wind turbines, microturbines, fuel cells |
| Loads | Typical residential needs: Lighting, Freezing, Refrigeration, Clothes Drying, Cooking, Water Heating, Space Cooling |
| Scalability | Yes |
| Communication | Yes, TCP/IP or Internet Protocol |
| Microgrids | One microgrid |
| Islanded | No, but it could be in emergency situations |
| Network type | Both |
| Models | Mathematical Formulae, Equations |
| Timescales | Milliseconds and hours |
| Commodity trade | Yes but adopting the "supply-driven-demand" approach instead of the traditional demand-driven-supply approach. |
| Dynamic prices | Yes |
| Time frames | Minute, Hour and Day. Electricity price ($/kWh) can vary from every minute to every hour. |
| | **Family Problem Dimension** |
| Supply-Driven-Demand Management | An IDAPS microgrid employs a bulletin board that represents an electricity market place for electricity buyers and sellers to transact business in. However, instead of matching the supply with the demand and producing market clearing prices, end-use customers make a decision, whether to buy electricity or not for their deferrable demand according to the real-time electricity pricing information offered by suppliers in the grid. |
| | **Societal Evaluation Criteria** |
| Environment | Not available |
| Economic | Not available |
| QoS | No real indicators, just follow grid's voltage and frequency. Response time within about half an electrical cycle (*i.e.* less than 0.008 second for a 60-Hz system) from detecting the fault to stabilizing the grid |
| Human | Done thanks to their "User Agent" that makes features of an IDAPS microgrid accessible to users. A user agent monitors voltage, current, active and reactive power consumption at each critical and non-critical load. It also allows users to control the status of loads based on priority pre-defined by a user. It retrieves real-time information and displays all relevant information on an electronic console on the end-user premises. This may include price ($/kWhr), quantity (kWhr) and duration (clock time) of electricity available for purchase, as well as electricity pricing information. |

As examples, only two articles proposed within the IDEaS Project will be evaluated below using our framework. The first article tackles a problem of demand-side management. The second article deals with virtual power plants.

Demand-Side Management: A decentralized demandside management model is proposed and evaluated in [48]. Such a model aims at optimizing the deferment of loads so as to maximize the comfort in the home and minimize energy cost. The proposed method can be evaluated on the basis of the criteria shown in **Table 4**.

**Table 4**. IDEaS project evaluation: demand-side management.

| | |
|---|---|
| | **Structural Dimension** |
| Physical | Yes |
| Size | Composed of 5000 homes, using load profiles of 26 M homes in the UK. |
| Storage | Not required for this kind of problem. |
| Sources | Not required for this kind of problem. |
| Loads | Deferrable loads, including shiftable static loads (like washing machines and dishwashers) and thermal loads (like fridges, boilers and radiators) and non-shiftable static loads (lighting and cooking). |
| Scalability | It has been shown in the article that the proposed approach can scale well |
| Communication | Surely but not described. |
| Microgrids | Represented by *smart homes*. |
| Islanded | Irrelevant for this kind of problem. |
| Network type | Simulated. |
| Models | Mathematical models are used for the loads. |
| Timescales | One decision per day and per agent. |
| Commodity trade | Demand-side management and pricing are dynamical and influence each other. |
| Dynamic prices | They are allocated dynamically, using the Real Time Pricing (RTP) based on the macro-model of the UK electricity market, but not between actors. |
| Time frames | Market prices are predicted for every next day over 100 experiments. |
| | **Family Problem Dimension** |
| Decentralized Demand-side Management | The proposed approach consists in coordinating large populations of agents. Each agent controls and adapts the deferrable loads of a *smart home* so as to maximize the comfort and minimize energy costs. The adaptive mechanism embedded in every agent is composed of two processes. First, a Widrow-Hoff learning mechanism is used to gradually adapt the deferment of the loads based on predicted market prices for the next day. Second, a stochastic mechanism is used to reoptimize the thermal load profiles on any particular day with a probability. |
| | **Societal Evaluation Criteria** |
| Environment | Minimizing carbon emissions. |
| Economic | Minimizing energy cost. |
| Quality of Service | Not taken into account. |
| Human Integration | Selecting which loads can be deferred. |

**Table 5**. IDEaS project evaluation: virtual plants.

| Structural Dimension | |
|---|---|
| Physical | Yes. |
| Size | The network has a nominal capacity of 17.5 MW and can supply 12000 homes. |
| Storage | Not mentioned. |
| Sources | 24 simulated wind turbines are used with profiles coming from real data of the Sotavento experimental wind farm in Spain. |
| Loads | Not available. |
| Scalability | No. |
| Communication | No. |
| Microgrids | Yes: the microgrids are represented by the cooperatives of VPP. |
| Islanded | Yes: only the wind turbines supply the demand. |
| Network type | Simulated. |
| Models | Analytical models are used. |
| Timescales | Slots of hours are used. |
| Commodity trade | Yes. |
| Dynamic prices | The prices are allocated dynamically between the cooperatives of VPP and the distribution network. |
| Time frames | Per hour. |
| **Family Problem Dimension** | |
| Virtual Power Plants | Virtual Power Plants represent collections of Distributed-Energy Resources (like storage systems or renewable-energy capacitor) that can be used more efficiently to meet load demand within an electricity supply network. Several techniques based on game theory are proposed in this article for forming and maintaining cooperative VPPs and for allowing them to reliably communicate their production to the electricity grid. |
| **Societal Evaluation Criteria** | |
| Environment | Not taken into account. |
| Economic | Not taken into account. |
| Quality of Service | Not taken into account. |
| Human Integration | Not taken into account. |

Virtual Power Plants: Although the production of energy with Distributed Energy Resources could potentially reduce reliance on conventional power plants, they lack the capacity, flexibility and controllability to participate in a cost-efficient way for demand supply, both in the physical electricity network and in the electricity market. The creation of Virtual Power Plants (VPP), that is entities that manage a set of DERs, has been suggested in recent years to cope with the previous quoted drawbacks. In [59], the emergence of cooperatives of VPP composed of small-to-medium size renewable electricity producers is controlled with a game-theoretic approach using a pricing mechanism and a scheme allocating payments within the cooperatives. **Table 5** evaluates the proposed technique using our framework.

## 5.5. PowerMatcher

The PowerMatcher [11] is one of the outcomes of the Smart House/Smart Grid European FP7 project[6]. It consists in a coordination mechanism that aims the balancing of demand and supply in (possibly multi-level) microgrids integrating Distributed Energy Resources. These microgrids are defined as clusters of sources and loads. The PowerMatcher mechanism implements supply and demand matching (SDM) using a multi-agent system and a market-based control approach. The PowerMatcher technology can be the basis of a Virtual Power Plant (VPP). The qualifying and evaluation of PowerMatcher is presented in table 6.

**Table 6**. Power matcher evaluation.

| | **Structural Dimension** |
|---|---|
| Physical | Yes |
| Size | Mass application (Hoogkerk, the Netherlands), mix of domestic clusters and selected grid segments (Mannheim, Germany), reaction to critical situations (Meltemi, Greece). |
| Storage | Electricity storage is present in the form of batteries. |
| Sources | Stochastic operation devices are present in the form of photovoltaic (PV) solar cells and small-scale wind turbines; classical sources are also present depending on the field test. |
| Loads | Deferrable loads, Shiftable static loads and classical loads depending on the field pilot. |
| Scalability | Yes it is one of the goals of the PowerMatcher. |
| Communication | The communication is implemented by a smart metering approach and via web services. |
| Microgrids | Yes the inherent structure defined by the PowerMatcher is hierarchical and allows to decompose a grid into smaller interacting grids in a VPP fashion. |
| Islanded | Yes, it was one of the constraints of the third field test. |
| Network type | Both simulated and real network were used. |
| Models | The appliances are simulated according to energy profiles drawn from well-known sources |
| Timescales | The simulator timescale/frequency is one second. |
| Commodity trade | Is the network for commodity trade taken into account? Yes/No |
| Dynamic prices | The prices are allocated dynamically, using the Real-Time Pricing (RTP) based on the macro-model of the UK electricity market, but not between actors. |
| Time frames | Market prices are predicted for every next day. |
| | **Family Problem Dimension** |
| SDM | The basic mechanism underlying PowerMatcher consists in answering the Supply and Demand Matching problem. The logical structure used also allows to tackle the VPP management problem. |
| | **Societal Evaluation Criteria** |
| Environment | Concerning the societal viewpoint RES and batteries have been used in pilots. Some initial results show that the use of PowerMatcher may bring from 14% to 21% of $CO_2$ reduction. |
| Economic | The peak reduction has been quantified at 4.49% (annual average) for the Greek pilot. |
| QoS | There is no evidence of the Quality of service issue in the available documents. |
| Human | The clients are integrated as real actors via the prosumer concept. Indeed, each house of the Netherlands pilot has a Home Energy Management (HEM) box, which implements the local energy management strategy of the house. |

Within a PowerMatcher cluster, the agents are organized into a logical tree. The leafs of this tree are a number of local device agents and, optionally, a unique objective agent. The root of the tree is formed by the auctioneer agent, a unique agent who handles the price forming, i.e., the search for the equilibrium price.

Follows a description of the three pilots.

- Netherlands

Twelve have a combined heat and power (CHP) plants with high efficient 1 kW boilers running on natural gas. The other 13 have a hybrid heat pump that combines an air-to-water heat pump with condensing boilers. All 25 houses have smart meters from Itron. Each house has twelve square meters of PV paneling for a total capacity of 1400 W peak. Ten houses have a smart washing machine and dishwasher from Miele. One house has a plugin hybrid Toyota Prius car and two others each has an all-electric Volkswagen Golf car. One house has a standard lead-acid battery to store solar energy for later use.

- Germany The installation of the smart meters has been done during spring 2010. The installation of first system prototypes at the end customers' premises have first started at "very friendly user's" homes. Afterwards, all BEMIs will be rolled out during October and November, so that the field trial operations can start. Meltemi comprises 220 cottages, which are fully inhabited in the summer (from May to September) and mostly empty in winter. A typical cottage in the camp is a single floor building of 70 $m^2$ surface. Most of the cottages are more than 30 years old.

- Greece The whole camp is supplied by a 3-phase medium-tolow voltage substation. The maximum load consumption of the site is approximately 220 kW. Furthermore, a 40 kW diesel generator is present. Finally, PV systems are installed in some of the houses as well in the entrance of the camping. The total installed capacity is 6 kWp.

# 6. CONCLUSIONS

This paper presents a framework for qualifying and evaluating smart grids approaches. This contribution is mainly based on a comparative study that is conducted based on an assessment framework analyzing each approach. The comparison is only possible for approaches that have same structure and same objective, thus the paper first provides focus on structural dimension and problem dimension to bring together approaches, then gives societal evaluation criteria in order to compare these different approaches.

This study aims at defining a cartography of the existing contributions to smart grid and analyze their strengths and weaknesses. Our objective was not to determine which approach is the best among the chosen ones. Such a choice would be dependent on many conditions specific to the deployment context. However, the survey presented in this paper should help a stakeholder with the comparison of the defined features. The reader should be aware of the following limits of this work:

- The results presented in this work for all chosen approaches are based on the available documentation (articles, technical reports and presentations). We may have missed some elements and could not be absolutely sure that the presented information is complete.
- The study is theoretical. There was no real experimentation to test the different approaches.
- The features defined for the analysis and comparison were influenced by the current state of the art of the domain which is still in its infancy. Future works may improve these features and go further.
- Future directions for this work may consist in deploying a website in order to store the presented results, enable new experiments and the addition of new features.

## REFERENCES

1. M. Suga, "Nedo's Smart Grid Related Activities," ADEME-NEDO Workshop, 2009.
2. SGMM Team, "Smart Grid Maturity Model," Technical Report CMU/SEI-2011-TR-025, 2011.
3. S. Pullins, "West Virginia Smart Grid Implementation Plan," Technical Report DOE/NETL-2009/1386, West Virginia Division of Energy, National Energy Technology Laboratory, US Department of Energy, 2009.

4. "European Smart Grids Technology Platform—Vision and Strategy for Europe's Electricity Networks of the Future," Technical Report, European Union, 2006.

5. Q.-N. N. Tran and G. C. Low. "Agent-Oriented Methodologies, Chapter XII: Comparison of Ten Agent-Oriented Methodologies, Pages 341-367," Idea Group, 2005.

6. S. D. J. McArthur, E. M. Davidson, V. M. Catterson, A. L. Dimeas, N. D. Hatziargyriou, F. Ponci and T. Funabashi, "Multi-Agent Systems for Power Engineering Applications—Part I: Concepts, Approaches, and Technical Challenges," IEEE Transactions on Power Systems, Vol. 22, No. 4, 2007, pp. 1743-1752.

7. S. D. J. McArthur, E. M. Davidson, V. M. Catterson, A. L. Dimeas, N. D. Hatziargyriou, F. Ponci and T. Funabashi, "Multi-Agent Systems for Power Engineering Applications—Part II: Technologies, Standards, and Tools for Building Multi-Agent Systems," IEEE Transactions on Power Systems, Vol. 22, No. 4, 2007, pp. 17531759.

8. K. Herter, T. O'Connor and L. Navarro, "A Systematic Approach for Assessing Plans to Benefit Customers and the Environment," Technical report, Herter Energy Research Solutions, Environmental Defense Fund, 2011.

9. S. J. Bossart and J. E. Bean, "Metrics and Benefits Analysis and Challenges for Smart Grid Field Projects," 2011 IEEE Energytech, Cleveland, 25-26 May 2011, pp. 1-5.

10. M. Wakefield, "Methodological Approach for Estimating the Benefits and Costs of Smart Grid Demonstration Projects," Technical Report, Electric Power Research Institute (EPRI), 2010.

11. J. K. Kok, M. J. J. Scheepers and I. G. Kamphuis, "Intelligence in Electricity Networks for Embedding Renewables and Distributed Generation," In R. R. Negenborn, Z. Lukszo and H. Hellendoorn, Eds., Intelligent Infrastructures, Vol. 42 of Intelligent Systems, Control and Automation: Science and Engineering, Springer, Dordrecht, 2010, pp. 179-209.

12. K. H. van Dam, "Capturing Socio-Technical Systems with Agent-Based Modeling," PhD Thesis, Delft University of Technology, Delft, 2009.

13. M. H. Albadi and E. F. El-Saadany, "Demand Response in Electricity Markets: An Overview," IEEE Power Engineering Society General Meeting, Tampa, 24-28 June 2007, pp. 1-5.

14. R. H. Lasseter and P. Paigi, "Microgrid: A Conceptual Solution," IEEE 35th Annual of Power Electronics Specialists Conference, Vol. 6, June 2004, pp. 4285-4290.

15. M. Amin, "Towards Self-Healing Energy Infrastructure Systems," IEEE Computer Applications in Power, Vol. 14, No. 1, 2001, pp. 20-28.

16. S. D. Ramchurn, P. Vytelingum, A. Rogers and N. R. Jennings, "Putting the 'Smarts' into the Smart Grid: A Grand Challenge for Artificial Intelligence," Communications of the ACM, Vol. 55, No. 4, 2012, pp. 86-97.

17. S. Salam, "Unit Commitment Solution Methods," World Academy of Science, Engineering and Technology 35, 2007.

18. C. W. Gellings and J. H. Chamberlin. "Demand-Side Management: Concepts and Methods," 2008.

19. W. M. Adams, "The Future of Sustainability: Re-Thinking Environment and Development in the Twenty-First Century," Report of the IUCN Renowned Thinkers Meetings, 29-31 January 2006.

20. National Renewable Energy Laboratory, "Power Technologies Energy Data Book," US Department of Energy, 2006.

21. J. T. Houghton and Intergovernmental Panel on Climate Change, "Climate Change 1995: The Science of Climate change. Climate Change 1995 [Intergovernmental Panel on Climate Change]. Cambridge University Press, Cambridge, 1996.

22. D. Weisser, "A Guide to Life-Cycle Greenhouse Gas (GHG) Emissions from Electric Supply Technologies," Energy, Vol. 32, No. 9, 2007, pp. 1543-1559.

23. International Energy Agency (IEA), "$CO_2$ Emissions from Fuel Combustion," OECD/IEA, Paris, 2011.

24. C. Harris, "Electricity Markets: Pricing, Structures and Economics," The Wiley Finance Series, John Wiley & Sons Incorporated, Hoboken, 2006.

25. "Energy Efficiency Indicators, a Study of Energy Efficiency Indicators for Industry in APEC Economies," Technical Report, Asia Pacific Energy Research Centre, Institute of Energy Economics, Tokyo, 2000.

26. ADEME, "Energy Efficiency Indicators: The European Experience," ADEME Editions, Paris, 1999.

27. V. Giordano, I. Onyeji, G. Fulli, M. S. Jiménez and C. Filiou, "Guidelines for Conducting Cost-Benefit Analysis of Smart Grid Projects," JRC Reference Reports, European Commission, 2012.

28. M. Sullivan and J. Schellenberg, "Smart Grid Economics: The Cost-Benefit Analysis," Renew Grid, 2011.

29. "Performance Incentive Scheme," Technical Report, Energy Regulatory Commission, 2005.

30. "IEEE Recommended Practice for Monitoring Electric Power Quality," IEEE Std 1159-2009 (Revision of IEEE Std 1159-1995), 2009, p. 26.

31. "5th CEER Benchmarking Report on the Quality of Electricity Supply 2011," Technical Report, Council of European Energy Regulators (CEER), 2012.

32. P. M. Johnson, "Human Centered Information Integration for the Smart Grid," Technical Report CSDL-09-15, Collaborative Software Development Laboratory, University of Hawaii at Manoa, 2009.

33. T. Voolink and R. Meertens, "The Effectiveness of Electronic Feedback on Energy and Hot Water Use by Householders," Technical Report, Maastricht University, 1999.

34. J. H. van Houwelingen and W. F. van Raaij, "The Effect of Goal-Setting and Daily Electronic Feedback on InHome Energy Use," The Journal of Consumer Research, Vol. 16, No. 1, 1989, pp. 98-105.

35. H. Staats, P. Harland and H. Wilke, "Effecting Durable change: A Team Approach to Improve Environmental Behavior in the Household," Environment and Behavior, Vol. 36, No. 3, 2004, pp. 341-367.

36. A. Faruqui, S. Sergici and A. Sharif, "The Impact of Informational Feedback on Energy Consumption: A Survey of the Experimental Evidence," Energy, Vol. 35, No. 4, 2009, pp. 1598-1608.

37. L. Becker, "Joint Effect of Feedback and Goal Setting on Performance: A Field Study of Residential Energy Conservation," Journal of Applied Psychology, Vol. 63, No. 4, 1978, pp. 428-433

38. D. A. Cohen, "Grid Agents: Intelligent Agent Applications for Integration of Distributed Energy Resources within Distribution Systems," IEEE Power and Energy Society General Meeting-Conversion and Delivery of Electrical Energy in the 21st Century, Pittsburgh, 20-24 July 2008, pp. 1-5.

39. G. James, D. Cohen, R. Dodier, G. Platt and D. Palmer, "A Deployed Multi-Agent Framework for Distributed Energy Applications," Proceedings of the Fifth International Joint Conference on Autonomous Agents and MultiAgent Systems, Hakodate, 8-12 May2006, pp. 676-678.

40. J. M. Akkermans, F. Ygge and R. Gustavsson, "Homebots: Intelligent Decentralized Services for Energy Management," 4th International Symposium on the Management of Industrial and Corporate Knowledge, Rotterdam, 21-22 October 1996, pp. 128-142.

41. F. Ygge, R. Gustavsson and H. Akkermans, "Homebots: Intelligent Agents for Decentralized Load Management," DA/DSM 96 Europe Distribution Automation & Demand Side Management Vienna, Vienna, 8-10 October 1996.

42. M. Pipattanasomporn, H. Feroze and S. Rahman. "MultiAgent Systems in a Distributed Smart Grid: Design and Implementation," IEEE Power Systems Conference and Exposition, Seattle, 15-18 March 2009, pp. 1-8.

43. H. S. Nwana, D. T. Ndumu, L. C. Lee and J. C. Collis, "Zeus: A Toolkit for Building Distributed Multiagent Systems," Applied Artificial Intelligence, Vol. 13, No. 1-2, 1999, pp. 129-185.

44. O. Tremblay, "A Generic Battery Model for the Dynamic Simulation of Hybrid Electric Vehicles," IEEE Vehicle Power and Propulsion Conference (VPPC), Arlington, 9- 12 September 2007, pp. 284-289.

45. A. Rogers, S. Maleki, S. Ghosh and N. R. Jennings, "Adaptive Home Heating Control through Gaussian Process Prediction and Mathematical Programming," 2nd International Workshop on Agent Technology for Energy Systems (ATES 2011), Taipei, 2-3 May 2011, pp. 71-78.

46. H. Rose, A. Rogers and E. H. Gerding, "Mechanism design for Aggregated Demand Prediction in the Smart Grid," AAAI Workshop on Artificial Intelligence and Smarter Living: The Conquest of Complexity, San Francisco, 7-8 August 2011.

47. H. Rose, A. Rogers and E. H. Gerding, "A Scoring Rule-Based Mechanism for Aggregate Demand Prediction in the Smart Grid," The 11th International Conference on Autonomous Agents and Multiagent Systems (AAMAS 2012), Valencia, 4-8 June 2012, pp. 661-668.

48. S. Ramchurn, P. Vytelingum, A. Rogers and N. Jennings, "Agent-Based Control for Decentralised Demand Side Management in the Smart Grid," The 10th International Conference on Autonomous Agents and Multiagent Systems (AAMAS 2011), Taipei, 2-6 May 2011, pp. 5-12.

49. S. Ramchurn, P. Vytelingum, A. Rogers and N. Jennings, "Agent-Based Homeostatic Control for Green Energy in the Smart Grid," ACM Transactions on Intelligent Systems and Technology, Vol. 2, No. 4, 2011, p. 35.

50. E. Gerding, V. Robu, S. Stein, D. Parkes, A. Rogers and N. Jennings, "Online Mechanism Design for Electric Vehicle Charging," 10th International Joint Conference on Autonomous Agents and Multi-Agent Systems (AAMAS), Taipei, 2-6 May 2011, pp. 811-818.

51. V. Robu, S. Stein, E. Gerding, D. Parkes, A. Rogers and N. Jennings, "An Online Mechanism for Multi-Speed Electric Vehicle Charging," 2nd International Conference on Auctions, Market Mechanisms and Their Applications (AMMA), New York, 22-23 August 2011, pp. 100-112.

52. S. Stein, E. Gerding, V. Robu and N. Jennings, "A Model-Based Online Mechanism with Pre-Commitment and Its Application to Electric Vehicle Charging," 11th International Joint Conference on Autonomous Agents and Multi-Agent Systems (AAMAS), Valencia, 4-8 June 2012, pp. 669-676.

53. P. Vytelingum, S. D. Ramchurn, T. D. Voice, A. Rogers, and N. R. Jennings, "Trading Agents for the Smart Electricity Grid," The 9th International Conference on Autonomous Agents and Multiagent Systems (AAMAS), Toronto, 9-14 May 2010, pp. 897-904.

54. S. Miller, S. D Ramchurn and A. Rogers, "Optimal Decentralised Dispatch of Embedded Generation in the Smart Grid," 11th

International Joint Conference on Autonomous Agents and Multi-Agent Systems (AAMAS), Valencia, 4-8 June 2012, pp. 281-288.

55. M. Alam, A. Rogers and S. Ramchurn, "A Negotiation Protocol for Multiple Interdependent Issues Negotiation over Energy Exchange," IJCAI Workshop on AI for an Intelligent Planet, Barcelona, 16-18 July 2011.

56. V. Robu, I. Vetsikas, E. Gerding and N. Jennings, "Addressing the Exposure Problem of Bidding Agents Using Flexibly Priced Options," 19th European Conference on Artificial Intelligence (ECAI), Lisbon, 16-20 August 2010, pp. 581-586.

57. "Agent-Based Micro-Storage Management for the Smart Grid," 9th International Joint Conference on Autonomous Agents and Multiagent Systems (AAMAS), Toronto, 9-14 May 2010, pp. 39-46.

58. T. Voice, P. Vytelingum, S. Ramchurn, A. Rogers and N. Jennings, "Decentralised Control of Micro-Storage in the Smart Grid," AAAI-11: 25th Conference on Artificial Intelligence, San Francisco, 7-11 August 2011, pp. 1421- 1426.

59. G. Chalkiadakis, V. Robu, R. Kota, A. Rogers and N. Jennings, "Cooperatives of Distributed Energy Resources for Efficient Virtual Power Plants," 10th International Joint Conference on Autonomous Agents and Multiagent Systems (AAMAS), Taipei, 2-6 May 2011, pp. 787-794.

# 3

# CHAPTER

# FLOCKING FOR LEADER-FOLLOWING MULTI-AGENT SYSTEMS WITH TIME-VARYING DELAY

**Fenglan Sun[1], Rui Wang[1], Yongfu Li[2], Feng Liu[3]**

[1]*Research Center of System Theory and Application, Chongqing University of Posts and Telecommunications, Chongqing, China*

[2]*College of Automation, Chongqing University of Posts and Telecommunications, Chongqing, China*

[3]*School of Automation, China University of Geosciences, Wuhan, China*

## ABSTRACT

This paper investigates the flocking problem in multi-agent system with time-varying delay and a virtual leader. Each agent here is subject to nonlinear dynamics. For the system, the corresponding algorithm with time-varying delay is proposed. Under the assumption that the initial network is connected, it is proved that the distance between agents is in the desired distance. The theore- tical deduction shows that the stable flocking motion is achieved.

**Keywords:** Flocking, Time-Varying Delay, Nonlinear Dynamics, Multi-Agent System, Virtual Leader

# 1. INTRODUCTION

Flocking is a collective behavior of large number of interacting agents with a common group objective. Examples of these agents include birds, fish, penguins, ants, bees, and crowds. Many scientists from rather diverse disciplines, including physics, mathematics, control engineering and biology, have been interested in flocking problem [1] -[8] . The first well-known flocking model was proposed by Craig Reynolds [2] . Reynolds started with a boid model to build a simulated flock and introduced three rules (i.e., separation, cohesion and alignment rules) for flocking. Based on Reynolds' three rules, flocking problems have been investigated from various perspectives [9] -[15] . In [16] , an artificial potential function is put forward and three algorithms are introduced. It provides a theoretical framework for the designing of flocking algorithms. Multi-agent flocking under topological interactions is considered, which define a notion of hierarchical structure in the interaction graph that establish conditions building upon previous work on multi-agent systems with switching communication networks in [17] . Using structurally balanced signed graph theory and a specified potential function, a stable bipartite flock formation is achieved for both virtual leader and leaderless situations in [18] . However, there is a common assumption that virtual leaders guide the flocking behaviors of the group. In this paper, leader follower flocking problem of multi-agent system is considered.

A flocking problem concerning multiple leaders in which followers use the position of flocking center to keep their connections is studied in [19] . In [20] , two leader-follower adaptive flocking algorithms are proposed with the combination of consensus and attraction/repulsion function respectively to solve the cohesive flocking problem and the formation flocking problem. Aiming at the group of autonomous agents consisting of multiple leader agents and multiple follower ones, a flocking behavior method with multiple leaders and a global trajectory was proposed in [21] . Yu et al., [22] give a distributed leader-follower algorithm considering the group consisting of one leader. In [23] , for the circumstance with a virtual leader, the agents would follow the virtual leader and achieve the same velocity asymptotically.

In practice, time delay is inevitable and would damage the stability of system. Jing et al. [23] investigate flocking problem of multi-agent systems with time delay and discuss systems with homogeneous and inhomogeneous time delay. Yang et al. [24] proposed an adaptive flocking algorithm for multi-agent system with time delay. It is proved that the distance between agents can be larger than a constant during the motion evolution by using the flocking algorithm. The authors investigate the flocking problem of multi-agent systems led by one active virtual leader with a directed topology containing time-varying coupling delays, which based only on the three classical assumptions for flocking systems in [25] . Because of these problems, this paper will study the flocking problems in the multi-agent system with a virtual leader and time-varying delay.

The rest of this paper is organized as follows. Some basic preliminaries and flocking algorithms are presented in Section 2. Section 3 gives the nonlinear leader-following multi-agent models. Algorithms and main results are presented in Section 4. Section 5 concludes the paper and offers suggestions for future work.

## 2. PRELIMINARIES

In this section, some related preliminary knowledge are introduced. For any vector $x$, the symbol $x^{\mathrm{T}}$ denotes its transpose and $\|x\|$ denotes the Euclidean norm. Let $\mathcal{G}=\{\mathcal{V},\mathcal{E},\mathcal{A}\}$ be a weighted undirected graph with the set of nodes $\mathcal{V}=\{1,2,\cdots,N\}$, and the set of agents $\mathcal{E}\subseteq\mathcal{V}\times\mathcal{V}$.

Graphs with self-loops will not be considered in this paper. The weight adjacency matrix $\mathcal{A}=(a_{ij})_{N\times N}$, where $a_{ij}\in(0,1]$ if $(i,j)\in\mathcal{E}$, otherwise, $a_{ij}=0$. An edge denoted by the pair $(j,i)$ represents a commu- nication link from node j to i. A path $\mathcal{H}$ from node i to node j is a sequence of edges, $\mathcal{H}=\{(i,k_1),(k_1,k_2),\cdots,(k_l,j)\}\subset\mathcal{E}$, in which all nodes $k_l, l=1,2,\cdots,l$ are distinct. An undirected graph is called connected if there is a path between each pair of distinct nodes. $D=\mathbf{diag}(d_1,d_2,\cdots,d_N)$ is the degree matrix whose diagonal elements are defined by $d_i=\sum_{j\neq i}a_{ij}$. The Laplacian matrix of graph $\mathcal{G}$ is $L=[l_{ij}]_{N\times N}=D-\mathcal{A}\in\mathcal{R}^{N\times N}$. Then it has following properties [26] ,

1) The eigenvalues of L satisfy $0=\lambda_1(L)\le\lambda_2(L)\le\cdots\le\lambda_N(L)$. If the graph $\mathcal{G}$ is connected, there is

$$\lambda_2(L)=\min_{z\perp 1_n}\frac{z^{\mathrm{T}}Lz}{\|z\|^2}>0.$$

2) The Laplacian matrix L is a positive semi-definite matrix that satisfies the following sum-of-squares property:

$$z^{\mathrm{T}}Lz=\frac{1}{2}\sum_{i,j\in\mathcal{E}}a_{ij}\left(z_i-z_j\right)^2,\ z\in R^n.$$

Lemma 1. [4] Suppose that $\mathcal{G}$ is an undirected graph of order N, and $\mathcal{G}_1$ is a graph generated by adding some edges into the graph $\mathcal{G}$. Then $\lambda_2(L(\mathcal{G}_1))\ge\lambda_2(L(\mathcal{G}))$, where $L(\mathcal{G})$ and $L(\mathcal{G}_1)$ are the symmetric Laplacian matrices of graphs $\mathcal{G}$ and $\mathcal{G}_1$, respectively.

Lemma 2. For any vectors $x,y\in R^n$, the following matrix inequality holds: $2x^{\mathrm{T}}y\le x^{\mathrm{T}}x+y^{\mathrm{T}}y$.

# 3. PROBLEMS FORMULATION

Consider the multi-agent system described by

$$\begin{cases}\dot{x}(t)=v_i(t),\\ \dot{v}_i(t)=f\left(v_i(t)\right)+u_i(t),\end{cases}\tag{1}$$

where $i=1,2,\cdots,N,x_i\in R^n,v_i\in R^n$ are the position and velocity states of i<sup>th</sup> agent, respectively. $f\left(v_i(t)\right)\in R^n$ is the nonlinear dynamic of agent i and $u_i\in R^n$ is the control input. Denote $x_{ij}=x_i-x_j,(i,j)\in\mathcal{E}$, as the relative distance between agent i and agent j.

For the systems with virtual leader available, the dynamics of virtual leader is described as

$$\begin{cases} \dot{x}_r(t) = \dot{v}_r(t), \\ \dot{v}_r(t) = f(v_r(t)), \end{cases} \tag{2}$$

where $q_r \in R^m, p_r \in R^m, u_r \in R^m$ represent the position, velocity and control input vector of the virtual leader.

Assumption (A): There exists a positive constant $\rho$ satisfying

$$\|f(z_1) - f(z_2)\| \leq \rho \|z_1 - z_2\|, \forall\, z_1, z_2 \in R^n. \tag{3}$$

Supposed that all agents have the same sensing radius $R > 0$. Then the neighboring set of agent i is denoted as $N_i(t) = \{j \mid \|x_i(t) - x_j(t)\| < R, (i,j) \in \mathcal{E}\}$.
Since the size of agent cannot be ignored usually that a minimum allowable distance r (collision distance) is considered in the model.

Definition 1: Given a constant $\varepsilon \in (0, R)$, $\mathcal{G}(t)$ is called a dynamic undirected graph with a time-varying set of links $\mathcal{E}(t) = \{(i,j) \mid i, j \in \mathcal{V}\}$ such that

1) Initial links are generated by $\mathcal{E}(0) = \{(i,j) \mid r < \|x_{ij}(0)\| < R\}$;

2) If $(i,j) \notin \mathcal{E}(0)$ and $\|x_{ij}(t)\| \leq R - \varepsilon$, then $(i,j)$ is a new link to be added $\mathcal{E}(t)$.
It is called hysteresis effect and $\gamma$ is the hysteresis distance, which is crucial in preserving connectivity of the network;

3) If $\|x_{ij}(t)\| \geq R$, then $(i,j) \notin \mathcal{E}(t)$.

The neighboring set of agent i is divided into four regions, named collision region, separation region, alignment region and attraction region, in which $R > R - \varepsilon > d_2 > d_1 > r$. If $d_1 < \|x_{ij}(t)\| < d_2$, agents i and j are in desired distance.

Definition 2. $\phi(\cdot)$ is a bounded function with respect to $z \in (r, R)$ between agents i and j, which satisfies

1) $\phi(z) < 0$ when $z \in (r, d_1)$;

2) $\phi(z)>0$ when $z\in(d_2,R)$;

3) $\phi(z)=0$ otherwise.

Definition 3. The pairwise bounded potential function $\psi(z):(0,\infty)\to(0,\infty)$ can be defined.

$$\psi(z)=\begin{cases}\int_{d_1}^{z}\phi(s)\mathrm{d}s, & z\in(0,d_1),\\ 0, & z\in[d_1,d_2],\\ \int_{d_2}^{z}\phi(s)\mathrm{d}s, & z\in(d_2,\infty),\end{cases} \tag{4}$$

which satisfies

1) $\psi(z)$ decreases with the increase of z when $z\in(0,d_1)$.

2) $\psi(z)$ is increasing on $(d_2,\infty)$. Obviously, the potential function $\psi(z)$ reaches its minimum value 0 when $z\in(d_1,d_2)$.

According to conditions, $\phi(z)$ can be constructed as follows:

$$\phi(\|x_{ij}\|)=\begin{cases}a_1(\|x_{ij}\|-d_1)\mathrm{e}^{\frac{(\|x_{ij}\|-d_1)^2}{b_1}}, & \|x_{ij}\|\in(r,d_1);\\ a_2(\|x_{ij}\|-d_2)\mathrm{e}^{\frac{(\|x_{ij}\|-d_2)^2}{b_2}}, & \|x_{ij}\|\in(d_2,R);\\ 0, & \text{otherwise.}\end{cases}$$

where the parameters $a_1,b_1,a_2,b_2$ are positive constants. There are $\psi(r)\geq Q_{\max},\psi(R)\geq Q_{\max}$, and $Q_{\max}=Q_0+(N(N-1)2-m_0)\psi(\|R-\varepsilon\|)$.

# 4. ALGORITHMS AND MAIN RESULTS

For system (1) with virtual leader (2), the flocking algorithm can be described by

$$u_i(t) = -\sum_{j\in N_i(t)} \phi(\|x_i(t)-x_j(t)\|) n_{ij}(t) - c_1\phi(\|x_i(t)-x_r(t)\|) n_{ir}(t) - c_2(v_i(t)-v_r(t))$$
$$-c\sum_{j\in N_i(t)} a_{ij}(v_i(t)-v_r(t)-v_j(t-\tau(t))+v_r(t-\tau(t))), \tag{5}$$

where $c, c_1, c_2$ are positive constants. Denote $\xi_i(t) = x_i(t) - x_r(t), \zeta_i(t) = v_i(t) - v_r(t)$, respectively. Then the system can be described by

$$\begin{cases} \dot{\xi}_i(t) = \zeta_i(t), \\ \dot{\zeta}_i(t) = f(v_i) - f(v_r) + u_i(t). \end{cases} \tag{6}$$

The control input (6) can be equivalently rewritten as

$$u_i(t) = -\sum_{j\in N_i(t)} \phi(\|\xi_i(t)-\xi_j(t)\|) n_{ij}(t) - c_1\phi(\|\xi_i(t)\|) n_{ir}(t) - c_2\zeta_i(t) - c\sum_{j\in N_i(t)} a_{ij}(\zeta_i(t)-\zeta_j(t-\tau(t))). \tag{7}$$

Denote $\xi(t) = [\xi_1^{\mathrm{T}}, \xi_2^{\mathrm{T}}, \cdots, \xi_N^{\mathrm{T}}]^{\mathrm{T}}, \zeta(t) = [\zeta_1^{\mathrm{T}}, \zeta_2^{\mathrm{T}}, \cdots, \zeta_N^{\mathrm{T}}]^{\mathrm{T}}$.

Definition 4. Flocking motion with a virtual leader is said to be achieved asymptotically for systems (1) and (2), if for any initial state, there is $d_1 \le \lim_{t\to\infty} \|x_i(t)-x_j(t)\| \le d_2,\ d_1 \le \lim_{t\to\infty} \|x_i(t)-x_r(t)\| \le d_2,\ \lim_{t\to\infty} \|v_i(t)-v_r(t)\| = 0,$ $\forall i = 1, 2, \cdots, N.$

To demonstrate the validity of control protocol (7), the following positive semi-definite function is constructed

$$Q = \frac{1}{2}\sum_{i=1}^{N}\sum_{j\in N_i}\psi(\|\xi_{ij}(t)\|) + \frac{1}{2}\sum_{i=1}^{N}\zeta_i^{\mathrm{T}}(t)\zeta_i(t) + \frac{1}{2}\sum_{i=1}^{N} c_1\int_{d_2}^{\xi_i}\phi(s)\,\mathrm{d}s + \sum_{i=1}^{N}\int_{t-\tau(t)}^{t}\zeta_i^{\mathrm{T}}(s)\zeta_i(s)\,\mathrm{d}s, \tag{8}$$

where $\xi_{ij}(t) = \xi_i(t) - \xi_j(t)$.

Theorem 1. Consider a multi-agent system modeled by dynamics (1) and (2), driven by control protocol (5). Suppose that the network is initially connected and $Q_0 := Q(\xi(0), \zeta(0))$ is bounded.When

$$\dot{\tau}(t) \le 1 - \frac{2\rho - c_2 - c\lambda_2 + 1 + \frac{1}{2}ck\lambda_N + \frac{1}{2}c\lambda_N}{k}$$, then following statements hold,

1) $G(t)$ is connected for all $t \geq 0$;

2) No collision occurs among agents for all $t \geq 0$;

3) Flocking motion with a virtual leader is achieved asymptotically.

Proof: Denote the topology switching time sequence as $t_k, k \in Z$. Without loss of generality, assume $t_0 = 0$. Taking the time derivative of the Lyapunov function Q on $t \in [t_0, t_1)$ gives

$$\begin{aligned}
\dot{Q} &= \sum_{i=1}^{N} \zeta_i^{\mathrm{T}}(t) \sum_{j \in N_i(t)} \phi(\|\xi_{ij}(t)\|) n_{ij}(t) + c_1 \sum_{i=1}^{N} \zeta_i^{\mathrm{T}}(t) \phi(\|\xi_i(t)\|) n_{ir}(t) + \sum_{i=1}^{N} \zeta_i^{\mathrm{T}}(t) \dot{\zeta}_i(t) \\
&\quad + \sum_{i=1}^{N} \left( \zeta_i^{\mathrm{T}}(t) \zeta_i(t) - (1 - \dot{\tau}(t)) \zeta_i^{\mathrm{T}}(t - \tau(t)) \zeta_i(t - \tau(t)) \right) \\
&= -\sum_{i=1}^{N} \left( \zeta_i^{\mathrm{T}}(t) \zeta_i(t) - (1 - \dot{\tau}(t)) \zeta_i^{\mathrm{T}}(t - \tau(t)) \zeta_i(t - \tau(t)) \right) \\
&\quad + \sum_{i=1}^{N} \zeta_i^{\mathrm{T}}(t) \Bigg( f(v_i) - f(v_r) - \sum_{j \in N_i(t)} \phi(\|\xi_i(t) - \xi_j(t)\|) n_{ij}(t) - c_1 \phi(\|\xi_i(t)\|) n_{ir}(t) \\
&\quad - c_2 \zeta_i(t) - c \sum_{j \in N_i(t)} a_{ij} \left( \zeta_i(t) - \zeta_j(t - \tau(t)) \right) \Bigg) \\
&\quad + \sum_{i=1}^{N} \left( \zeta_i^{\mathrm{T}}(t) \zeta_i(t) - (1 - \dot{\tau}(t)) \zeta_i^{\mathrm{T}}(t - \tau(t)) \zeta_i(t - \tau(t)) \right).
\end{aligned}$$

From Lemma 2, there is

$$\begin{aligned}
\dot{Q} &\leq 2\rho \|\zeta(t)\|^2 - c_2 \|\zeta(t)\|^2 - c\lambda_2(L) \|\zeta(t)\|^2 + \frac{1}{2} c\lambda_N(L) \|\zeta(t)\|^2 \\
&\quad + \frac{1}{2} c\lambda_N(L) \|\zeta(t - \tau(t))\|^2 + \|\zeta\|^2 - (1 - \dot{\tau}(t)) \|\zeta(t - \tau(t))\|^2.
\end{aligned}$$

For a positive constant k, one has $\|\zeta(t - \tau(t))\|^2 \leq k \|\zeta(t)\|^2$, thus

$$\begin{aligned}
\dot{Q} &\leq 2\rho \|\zeta(t)\|^2 - c_2 \|\zeta(t)\|^2 - c\lambda_2(L) \|\zeta(t)\|^2 + \frac{1}{2} c\lambda_N(L) \|\zeta(t)\|^2 \\
&\quad + \frac{1}{2} ck\lambda_N(L) \|\zeta(t)\|^2 + \|\zeta(t)\|^2 - (1 - \dot{\tau}(t)) \|\zeta(t - \tau(t))\|^2.
\end{aligned}$$

Assume that $\tau(t)$ satisfy $\dot{\tau}(t) \leq 1 - \dfrac{2\rho - c_2 - c\lambda_2 + 1 + \frac{1}{2} ck\lambda_N + \frac{1}{2} c\lambda_N}{k}$, there is

$\dot{Q} \le 0, \forall t \in [t_0, t_1)$,

which implies that

$$Q(t) \le Q_0 < \infty, \forall t \in [t_0, t_1).$$

By definition (2), one has $\psi(\|R\|) \ge Q_{\max}$. Therefore, no edge distance will be tend to R for $\forall t \in [t_0, t_1)$, implying that no existing edges will be lost before time $t_1$. Hence, new edges must be added into the network at $t_1$. For a system consists of N agents, there are at most $N(N-1)/2$ edges. At the initial instant $t_0$, the system consists of $m_0$ edges, then

$Q(t_k) \le Q_0 + (N(N-1)/2 - m_0)\psi(\|R-\varepsilon\|) = Q_{\max}$.

Hence there is no edge lost. In addition, from the definite of potential function, one has $\psi(\|r\|) \ge Q_{\max}$. Therefore, no collision occurs during $[t_0, t_1)$.

Similar to the above analysis, taking the time derivative of $Q(t)$ on every $[t_{k-1}, t_k)$. By lemma 1, there is

$$\dot{r}(t) \le 1 - \frac{2\rho - c_2 - c\lambda_2(L(0)) + 1 + c\lambda_N(L(0))k}{k} \le 1 - \frac{2\rho - c_2 - c\lambda_2(L(t)) + 1 + c\lambda_N(L(t))k}{k},$$

one has $\dot{Q}(t) \le 0$, which implies $Q(t) \le Q(t_{k-1}) < \infty, \forall t \in [t_{k-1}, t_k)$. Thus no edge distance will tend to R for $[t_{k-1}, t_k)$, implying that no edge will be lost before time $t_k$ and $Q(t_k)$ is finite. Since $\mathcal{G}(0)$ is connected and no edge in $E(0)$ is lost, $\mathcal{G}(t)$ will be connected for all $t \ge 0$. This completes the proof of part (1).

Similarly, from $\psi(\|r\|) \ge Q_{\max}$, deducing that no edge distance will tend to r, for all $t \ge 0$. Thus collision is avoided during the whole process. This completes the proof of part (2). To proof part (3), assume that there are $m_k$ new edges being added to the evolving network at time $t_k$. As no edges are lost for $t \geq 0$, and $m_k \leq N(N-1)/2 - m_0 \leq (N-1)(N-2)/2$. Therefore, the number of switching times of the system (1) is finite, which implies that the evolving network $\mathcal{G}(t)$ eventually becomes fixed. Denote the last topology switching as $t_k$. Then Q is continuous and monotonously decreasing for $t \in [t_k, \infty)$. Hence the set $\Omega := \{\xi \in D, \zeta \in R^{Nn} | Q(\xi, \zeta) \leq Q^*\}$, is positively invariant, where

$$D = \left\{\xi \in R^{N^2 n} \middle| \|\xi_{ij}\| \in \left[\min\left(\psi^{-1}(Q_{\max})\right), \max\left(\psi^{-1}(Q_{\max})\right)\right], (i,j) \in \mathcal{E}(t), t \in [t_k, \infty)\right\},$$

and $\xi = (\xi_{11}, \cdots, \xi_{1N}, \cdots, \xi_{N1}, \cdots, \xi_{NN})^{\mathrm{T}} \in R^{Nn}$.

Since $\mathcal{G}(t)$ is connected for all $t \geq 0$, one has $\|\xi_{ij}\| \leq (N-1)R$, for all $(i,j) \in \mathcal{E}(t)$. As $Q(t) \leq Q_{\max}$, one has $\zeta_i^{\mathrm{T}} \zeta_i \leq 2Q_{\max}$, thus $\|\zeta_i(t)\| \leq \sqrt{2Q_{\max}}$. Therefore, the set Ω is compact. It follows from LaSalle's invariance principle that if the initial condition lies in Ω, then the corresponding trajectorics will converge to the largest invariant set inside the region

$$S = \left\{\xi \in R^{N^2 n}, \zeta \in R^{Nn} \middle| \dot{Q}(\xi, \zeta) = 0\right\}.$$

From (8), $\dot{Q} = 0$ if and only if $v_1 = \cdots = v_N = v_r$, which implies that the velocities of all agents will converge to that of the virtual leader asymptotically.

Since $v_1 = \cdots = v_N = v_r$, there is $\dot{\zeta}_i = 0$ for all $i = 1, 2, \cdots, N$. From (6), one has

$$\begin{aligned}\dot{\zeta}_i(t) &= f(v_i) - f(v_r) + u_i(t) \\ &= -\sum_{j \in N_i(t)} \phi(\|x_i(t) - x_j(t)\|) n_{ij}(t) - c_1 \phi(\|x_i(t) - x_r(t)\|) n_{ir}(t) \\ &= 0.\end{aligned}$$

Thus, unless the inital configuration of the agents is close enough to the global minimum, almost every final configuration locally minimizes each agent's global potential. which implies

$$\|x_i - x_j\| \in [d_1, d_2];$$
$$\|x_i - x_r\| \in [d_1, d_2].$$

Then the flocking is achieved. This completes the proof of part (3), thus Theorem 1 hold.

Remark 1. If $\tau(t) = \tau$ is the constant delay, from the deduction above, Theorem 1 is also hold.

# 5. CONCLUSION

This paper mainly discusses the flocking problem of multi-agent system with a virtual leader and time-varying delay. Unlike most existing flocking algorithms, each agent here is subject to nonlinear dynamics. The corresponding algorithms with the time-varying delay are proposed. Under the assumption that the initial network is connected, the theoretical deduction is made. The related topic over the directed network or the jointly connected network will be studied in future.

# ACKNOWLEDGEMENTS

We thank the Editor and the referee for their comments. This work was supported by the National Nature Sci- ence Foundation of China under Grants 61503053, 61472374 and 61304197, the Natural Science Foundation Pro- ject of CQ CSTC, China (Grant No. cstc2013jcyjA40018), the Youth Science Research Project of CQUPT, China (Grant Nos. A2012-78 and A2012-82), the Doctor Start-up Foundation of CQUPT, China (Grant Nos. A2012-23 and A2012-26), the Natural Science Fundation of CQJW, China (Grant Nos. KJ130506 and KJ1400435), and Training Programme Foundation for the Talents of Higher Education Commission. This support is greatly appreciated.

## CITE THIS PAPER

FenglanSun,RuiWang,YongfuLi,FengLiu, (2016) Flocking for Leader-Following Multi-Agent Systems with Time-Varying Delay. *Intelligent Control and Automation*,**07**,9-15.

## REFERENCES


1. Toner, J. and Tu, Y. (1998) Flocks, Herds, and Schools: A Quantitative Theory of Flocking. Physical Review E, 58, 4828.
2. Reynolds, C.W. (1987) Flocks, Herds and Schools: A Distributed Behavioral Model. ACM Siggraph Computer Graphics, 21, 25-34.
3. Toner, J. and Tu, Y. (2005) Hydrodynamics and Phases of Flocks. Annals of Physics, 318, 170-244.
4. Olfati-Saber, R. (2006) Flocking for Multi-Agents with a Virtual Leader. IEEE Transactions on Automatic Control, 51, 410-420.
5. Crowther, B. (2002) Flocking of Autonomous Unmanned Air Vehicles. Aeronautical Journal, 107, 99-109.
6. Akyildiz, I.F., Su, W. and Cayirci, E. (2002) A Survey on Sensor Networks. IEEE Communications Magazine, 40, 102-114.
7. Partridge, B.L. (1984) The Chorus-Line Hypothesis of Maneuver in Avian Flocks. Nature, 309, 344-345.
8. Gazi, V. and Passino, K.M. (2004) Stability Analysis of Social Foraging Swarms. IEEE Transactions on Systems, Man, and Cybernetics, Part B: Cybernetics, 34, 539-557.
9. Ren, W. and Atkins, E. (2007) Distributed Multi-Vehicle Coordinated Control via Local Information Exchange. International Journal of Robust and Nonlinear Control, 17, 1002-1033.
10. Tanner, H.G., Jadbabaie, A. and Pappas, G.J. (2007) Flocking in Fixed and Switching Networks. IEEE Transactions on Automatic Control, 52, 863-868.

11. Su, H., Wang, X. and Lin, Z. (2009) Flocking of Multi-Agents with a Virtual Leader. IEEE Transactions on Automatic Control, 54, 293-307.

12. Topaz, C.M. and Bertozzi, A.L. (2004) Swarming Patterns in a Two-Dimensional Kinematic Model for Biological Groups. SIAM Journal on Applied Mathematics, 65, 152-174.

13. Chen, Z. and Zhang, H.T. (2013) A Remark on Collective Circular Motion of Heterogeneous Multi-Agents. Automatica, 49, 1236-1241.

14. Zhang, H.T., Zhai, C. and Chen, Z. (2011) A General Alignment Repulsion Algorithm for Flocking of Multi-Agent Systems. IEEE Transactions on Automatic Control, 56, 430-445.

15. Lu, X.Q., Austin, F. and Chen, S.H. (2011) Flocking in Multi-Agent Systems with Active Virtual Leader and Time-Varying Delays Coupling. Communications in Nonlinear Science and Numerical Simulation, 16, 1014-1026.

16. Olfati-Saber, R. (2006) Flocking for Multi-Agent Dynamic Systems: Algorithms and Theory. IEEE Transactions on Automatic Control, 51, 401-420.

17. Martin, S. (2014) Multi-Agent Flocking under Topological Interactions. Systems and Control Letters, 69, 53-61.

18. Fan, M.C., Zhang, H.T. and Wang, M. (2014) Bipartite Flocking for Multi-Agent Systems. Communications in Nonlinear Science and Numerical Simulation, 19, 3313-3322.

19. Gu, D. and Wang, Z. (2009) Leader-Follower Flocking: Algorithms and Experiments. IEEE Transactions on Control Systems Technology, 17, 1211-1219.

20. Jia, Y. and Zhang, W. (2014) Distributed Adaptive Flocking of Robotic Fish System with a Leader of Bounded Unknown Input. International Journal of Control, Automation and Systems, 12, 1049-1058.

21. Li, M., Liang, J. and Li, S. (2014) Flocking Behavior with Multiple Leaders and Global Trajectory. Journal of Central South University, 21, 2324-2333.

22. Yu, W., Chen, G. and Cao, M. (2010) Distributed Leader-Follower Flocking Control for Multi-Agent Dynamical Systems with Time-Varying Velocities. Systems & Control Letters, 59, 543-552.

23. Luo, J., Zhang, S., Kang, H., et al. (2013) Flocking Algorithms for Multi-Agent Systems with Time-Delay. Proceedings of the IEEE International Workshop on Microwave and Millimeter Wave Circuits and System Technology (MMWCST), Chengdu, 24-25 October 2013, 428-431.

24. Yang, Z., Zhang, Q. and Chen, Z. (2013) A Novel Adaptive Flocking Algorithm for Multi-Agents System with Time Delay and Nonlinear Dynamics. Proceedings of the 32nd IEEE Chinese Control Conference (CCC), Xi'an, 26-28 July 2013, 998-1001.

25. Lu, X., Austin, F. and Chen, S. (2011) Flocking in Multi-Agent Systems with Active Virtual Leader and Time-Varying Delays Coupling. Communications in Nonlinear Science and Numerical Simulation, 16, 1014-1026.

26. 26. Shi, H., Wang, L. and Chu, T. (2006) Virtual Leader Approach to Coordinated Control of Multiple Mobile Agents with Asymmetric Interactions. Physica D: Nonlinear Phenomena, 213, 51-65.

# 4

# CHAPTER

# PERTURBATION OF FRACTIONAL MULTI-AGENT SYSTEMS IN CLOUD ENTROPY COMPUTING

**Rabha W. Ibrahim *, Hamid A. Jalab and Abdullah Gani**

*Faculty of Computer Science and Information Technology, University of Malaya, Kuala Lumpur 50603, Malaysia*

## ABSTRACT

A perturbed multi-agent system is a scheme self-possessed of multiple networking agents within a location. This scheme can be used to discuss problems that are impossible or difficult for a specific agent to solve. Intelligence cloud entropy management systems involve functions, methods, procedural approaches, and algorithms. In this study, we introduce a new perturbed algorithm based on the fractional Poisson process. The discrete dynamics are suggested by using fractional entropy and fractional type Tsallis entropy. Moreover, we study the algorithm stability.

**Keywords:** fractional calculus; fractional differential equation; cloude entropy system; complexity

# 1. INTRODUCTION

Cloud entropy systems are currently increasing in number and developing. The increasing occurrence of clouds has led to a large number of unresolved tasks. Cloud computing recognizes the internet services delivered by generalized data centers in the form of hardware and software systems; it is also a distributed computing system that consists of a group of organized virtual machines [1]. The original technology in a cloud entropy system consists of long-established values of routing, visualization, code duplication, and so on. Overall, the system uses known computational methods. Recent network structure schemes involve the improvement of new leader lines for the future of computing procedures. Such networks present understandable improvements, such as applying feedback. This usage leads to the perturbation of cloud systems.

Another problem of cloud entropy systems is how to divide information without performance degradation. The effectiveness of standing transmission systems and databases in clouds should be analyzed. Developing measures for the system transmission efficiency is necessary to achieve improvements. A drawback system functioning in clouds is devoted to communication channels. The utility of perturbation offers the possibility of transfering to the dynamic control of a model, principles established for them, and conforming optimization methods. Control systems are applied to assess structural complexity because they have the ability to oversee all functioning units, the counting frequency through networks, and the present form of resources.

Multi-agent systems fundamentally appear in integer calculus by using ordinary differential equations. Various multi-agent systems cannot be repeated with the integer cases in a complex physical situation, but they are refined with fractional order (non-integer) differential equations [2]. Fractional multi-agent systems are considered and suggested by a collection of fractional differential operators, such as Riemann–Liouville, Caputo, and Jumarie [3,4,5]. Multi-agent systems have substantial documented research in the fields of military studies, economic systems, and control systems. An enormous number of these systems has been examined, such as distributed sensor networks, cloud computing systems composed of multiple servers, altitude control of satellites, and so on [6,7,8,9,10]. A perturbed multi-agent system is a scheme self-possessed of multiple networking agents within a location. This scheme can be utilized to address problems that are impossible or difficult for a specific agent to solve. Intelligent cloud-entropy management systems involve various functions, methods, procedural approaches, and algorithms. Overall, a dynamic system is necessary to improve a stable structure and its reliability. The decomposition of functional cloud components is conducted through certain methods for the system analysis of structural complexity. The dynamic explanations in the formal approximations of differential equations with control are employed in many works [11,12,13].

In this study, we establish a new perturbed multi-agent system for cloud entropy services based on the fractional Poisson process (see Figure 1). The discrete dynamics are proposed by using fractional entropy and fractional Tsallis entropy. Moreover, we examine the algorithm stability.

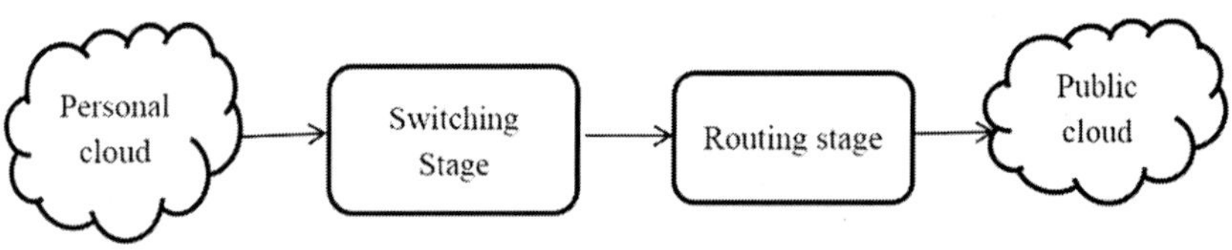


**Figure 1.** Structure of a perturbation cloud entropy system.

## 2. SETTING

In this section, we consider the basic positions in which incidence establishes grounds for the control system theory and dynamic models of cloud service. We reflect a distributed database in clouds. Cloud databases comprise software or hardware routing. They supply data on the practical machine on which sections of data are deposited, the private cloud that determines the level of security in public, and other service information. A similar type of entering query is transformed in each incident to different situations and route maps and is influenced by the model structure and communication frequencies.

In such a setting, a continuum of agents that have nonhomogenous preferences pay a cost to move from one point to another in the state space. We suppose that n agents exist, whose communication network is a directed graph. Agent i has cost function Φi(t,χ,u), which represents what an agent pays to have characteristics χi(t)(*i.e.*, the level of cloud computing at time t) under controller input ui(t). Thus, we may present the following fractional system:

$$D^{(\wp)}\chi_i(t) = \Phi_i(t,\chi,u), i = 1,...,n$$

$$(y_i(t) = \chi_i(t)),$$

where yi(t) is the outcome at time t, D(℘)χi(t) indicates the Riemann–Liouville calculus given by:

$$D^{(\wp)}\chi_i(t) = \frac{d}{dt}\int_a^t \frac{(t-\tau)^{-\wp}}{\Gamma(1-\wp)}\chi(\tau)\,d\tau,$$

which corresponds to the integral operator:

$$I_a^{\wp}\chi(t) = \int_a^t \frac{(t-\tau)^{\wp-1}}{\Gamma(\wp)}\chi(\tau)\,d\tau,$$

accordingly, $\wp \in (0,1)$. Saber *et al.* [14] presented the dynamics of a cloud through the following equation:

$$\chi_i'(t) = \Phi_i(t,\chi,u), i = 1,...,n$$
$$= \sum_{j\in\mathbb{N}} \alpha_{ij}(\chi_j(t) - \chi_i(t)),$$

where $\alpha_{ij}>0$ is the (i,j) (agent i can receive information from agent j; else, αij = 0) element of the adjacency matrix $\chi(t) = (\chi_1(t),...,\chi_n(t))^{\top}, y(t) = (y_1(t),...,y_n(t))^{\top}, u(t) = (u_1(t),...,u_n(t))^{\top}$ then the system is asymptotically stable (see [14]).

Our discussion is based on the linear perturbation system of the second type [15], that is:

$$D^{(\wp)}\left[\chi_i(t) + \phi_i(t,\chi,u)\right] = \varphi_i(t,\chi,u), i = 1,\ldots,n$$

$$(y_i(t) := \chi_i(t)),$$

(1)

where $\phi i(t,\chi i,ui)=\varphi i(t,\chi,ui)-D(\wp)\,\phi i(t,\chi,u)$.

Function $\chi i \in C[J,R]$, which is the space of all continuous functions on J, is called a solution for problem (1) if and only if function $\chi i$ is continuous for all $t \in J$ and $\chi i$ fulfills Equation (1). Our objective is to establish the existence of a solution for Equation (1). The main significant benefit of the construction usage of fractional differential equations (ordinary and partial) in mathematical modeling is their nonlocal property. The integer order differential operator is a local, linear operator, whereas the fractional order differential operator is a nonlocal, nonlinear one. Thus, the following state of a system is influenced by its present and past states (this property is useful in cloud computing). Thus, fractional calculus has become increasingly popular in technical and industrial fields. In this study, a perturbation analysis model is studied to solve linear and nonlinear differential and integral equations. Various types of perturbation techniques and forms are applied depending on the actual situation and physical problems. A good outcome is obtained in a cloud entropy system by minimizing the time of utility and the cost of the system (see [16]).

## 3. EXISTENCE RESULTS OF A PERTURBED SYSTEM

We define a supremum norm $\|\cdot\| \in C[J,\mathbb{R}]$ by $\|x\| := \sup_{t \in J}|x|$. Obviously, $C[J,\mathbb{R}]$ is a Banach space. We generate the following result, which can be found in [17]:

### Lemma 3.1

*We let* B *be a closed convex and bounded subset of the Banach space* X *and* P:X→X *and* Q:B→X*be two operators, such that (a)* P *is nonlinear D-contraction, (b)* Q *is compact and continuous, and (c)* $\chi$ = P$\chi$+Qy*for all* $\chi$,y∈B. *The operator equation* P$\chi$+Q$\chi$ = $\chi$ *then has a solution in* B.

By applying certain properties of fractional operators, we obtain the following result:

## Lemma 3.2

*We assume that* t⟼χi−ϕi *is injective (or increasing) in* R *and* φi *is bounded on* J×R×R. *Therefore, function* χi *is a solution of problem (1) if and only if it is a solution of the fractional integral equation*:

$$\chi_i(t) = \chi_i(t_0) - \phi_i(t_0, \chi(t_0), u(t_0)) + \phi_i(t, \chi, u) + \int_0^t \frac{(t-\tau)^{\wp-1}}{\Gamma(\wp)} \varphi_i(\tau, \chi(\tau), u(\tau))\, d\tau.$$

## Theorem 3.1

*We suppose that the assumptions of Lemma 3.2 hold. If constant* ℓ>0 *exists, such that*:

$$|\phi_i(t, \chi, u) - \phi_i(t, x, v)| \leqslant \ell\left(\frac{|\chi - x| + |u - v|}{\flat + |\chi - x| + |u - v|}\right), \ell \leqslant \flat, |\varphi_i| \leqslant \flat \quad (2)$$

*then problem (1) has a solution on* J.

## Proof of Theorem 3.1.

In view of Lemma 3.2, we have:

$$\begin{aligned}
|\chi_i| &= \left|\chi_i(t_0) - \phi_i(t_0, \chi(t_0), u(t_0)) + \phi_i(t, \chi, u) + \int_0^t \frac{(t-\tau)^{\wp-1}}{\Gamma(\wp)} \varphi_i(\tau, \chi(\tau), u(\tau))\, d\tau\right| \\
&\leqslant |\chi_0| + \ell\left(\frac{|\chi - \chi_0| + |u - u_0|}{\flat + |\chi - \chi_0| + |u - u_0|}\right) + \int_0^t \frac{(t-\tau)^{\wp-1}}{\Gamma(\wp)} |\varphi_i(\tau, \chi(\tau), u(\tau))|\, d\tau \\
&\leqslant |\chi_0| + \ell\left(\frac{|\chi - \chi_0| + |u - u_0|}{\flat + |\chi - \chi_0| + |u - u_0|}\right) + \flat \int_0^t \frac{(t-\tau)^{\wp-1}}{\Gamma(\wp)} d\tau \\
&= |\chi_0| + \ell\left(\frac{|\chi - \chi_0| + |u - u_0|}{\flat + |\chi - \chi_0| + |u - u_0|}\right) + \frac{\flat T^{\wp}}{\Gamma(\wp + 1)} := \mathcal{B}.
\end{aligned}$$

B is clearly a closed, convex, and bounded subset of X. We define the two operators *P:X⟶X* and *Q:B⟶X* as follows:

$$(P)(\chi) := \phi_i(t,\chi,u), (Q)(\chi) := \chi_0 - \phi_i(t,\chi_0,u_0) + \int_0^t \frac{(t-\tau)^{\wp-1}}{\Gamma(\wp)} \varphi_i(\tau,\chi(\tau),u(\tau))\,d\tau.$$

Thus, we have the following operator equation:

$$(\chi)(t) = (P)(\chi)(t) + (Q)(\chi)(t) \tag{3}$$

We aim to achieve the conditions of Lemma 3.1. First, we show that P is a nonlinear D-contraction on X. Using (2), we obtain:

$$|(P)(\chi,u)(t) - (P)(x,v)(t)| = |\phi_i(t,\chi,u) - \phi_i(t,x,v)|$$
$$\leqslant \ell(\frac{|\chi - x| + |u - v|}{\beth + |\chi - x| + |u - v|}),$$

By taking a supremum norm over J, we have:

$$\| (P)(\chi) - (P)(x) \| \leqslant \ell(\frac{\| \chi - x \| + \| u - v \|}{\beth + \| \chi - x \| + \| u - v \|}).$$

Thus, P is a D-contraction mapping defined by:

$$D(\rho) := \frac{\ell\rho}{\beth + \rho}, \rho = \| \chi - x \| + \| u - v \| .$$

Second, we aim to show that Q is a compact and continuous operator on B into X. To prove that Q is continuous on B, we let {χm},{um} be two sequences in B converging to point χ,u∈B, respectively. By utilizing the dominated convergence theorem for integration, we have:

$$\lim_{m\to\infty} Q(\chi_m)(t) = \lim_{m\to\infty} [\chi_0 - \phi_i(t,\chi_0,u_0) + \int_0^t \frac{(t-\tau)^{\wp-1}}{\Gamma(\wp)} \varphi_i(\tau,\chi_m(\tau),u_m(\tau))\,d\tau]$$
$$= \chi_0 - \phi_i(t,\chi_0,u_0) + \lim_{m\to\infty} [\int_0^t \frac{(t-\tau)^{\wp-1}}{\Gamma(\wp)} \varphi_i(\tau,\chi_m(\tau),u_m(\tau))\,d\tau]$$
$$= \chi_0 - \phi_i(t,\chi_0,u_0) + \lim_{m\to\infty} [\int_0^t \frac{(t-\tau)^{\wp-1}}{\Gamma(\wp)} \varphi_i(\tau,\chi(\tau),u(\tau))\,d\tau]$$
$$= (Q)(\chi)(t).$$

Q is accordingly continuous on B into X. We proceed to show that Q is uniformly bounded:

$$|Q(\chi)(t)| \leq |\chi_0 - \phi_i(t,\chi_0,u_0)| + [\int_0^t \frac{(t-\tau)^{\wp-1}}{\Gamma(\wp)} |\varphi_i(\tau,\chi(\tau),u(\tau))| d\tau]$$
$$\leq |\chi_0 - \phi_i(t,\chi_0,u_0)| + \frac{\flat T^{\wp}}{\Gamma(\wp+1)}.$$

Q is then proven to be uniformly bounded on B. For t2≥t1, t1,t2∈J, we obtain:

$$|Q(\chi)(t_1) - Q(\chi)(t_2)| = |\int_0^{t_1} \frac{(t_1-\tau)^{\wp-1}}{\Gamma(\wp)} \varphi_i(\tau,\chi(\tau),u(\tau))\,d\tau| - \int_0^{t_2} \frac{(t_2-\tau)^{\wp-1}}{\Gamma(\wp)} \varphi_i(\tau,\chi(\tau),u(\tau))\,d\tau||$$
$$\leq |\int_0^{t_2} \frac{(t_2-\tau)^{\wp-1}}{\Gamma(\wp)} |\varphi_i(\tau,\chi(\tau),u(\tau))|\,d\tau|$$
$$\leq \frac{(t_2-t_1)^{\wp}\flat}{\Gamma(\wp+1)} := \epsilon,$$

where |t2−t1|: = δ. Q(B) then becomes an equi-continuous set in X. Thus, by virtue of the Arzelá–Ascoli theorem, we conclude that Q is compact. To achieve condition (c), we have:

$$|\chi| = |P(\chi) + Q(\chi)|$$
$$\leq |\chi_0| + \ell(\frac{|\chi-\chi_0|+|u-u_0|}{\flat+|\chi-\chi_0|+|u-u_0|}) + \int_0^t \frac{(t-\tau)^{\wp-1}}{\Gamma(\wp)} |\varphi_i(\tau,\chi(\tau),u(\tau))| d\tau$$
$$\leq |\chi_0| + \ell(\frac{|\chi-\chi_0|+|u-u_0|}{\flat+|\chi-\chi_0|+|u-u_0|}) + \flat\int_0^t \frac{(t-\tau)^{\wp-1}}{\Gamma(\wp)} d\tau$$
$$= |\chi_0| + \ell(\frac{|\chi-\chi_0|+|u-u_0|}{\flat+|\chi-\chi_0|+|u-u_0|}) + \frac{\flat T^{\wp}}{\Gamma(\wp+1)}$$

Therefore, χ∈B. All the hypotheses of Lemma 3.1 are then demostrated; thus, the operator equation P+Q = χ has a solution in B. As a result, system (1) has a solution defined on J. This outcome completes the proof. □

## Remark 3.1

*For the special case in different forms, the equation:*

$$\chi(t+1) = P\chi(t) + Qu(t)$$

*was considered by Pluzhnik et al. [18]. They requested control* u *under constraint* u∈B⊆X. *Theorem 3.1 shows that the constraint in [18] is valid. In the following section, we illustrate an algorithm to minimize the cost function by utilizing the fractional Poisson process incorporating the fractional Tsallis entropy.*

# 4. THE PROPOSED ALGORITHM

To solve Equation (1), we need the following facts:

**Objective function**. Our primary objective is to investigate the optimal flat rate at which a company uses cloud computing. With the cost proposed in Section 2, all rational companies decide on the flat rate of switching to the cloud-computing pattern to minimize the estimated discount cost with respect to the effort cost. For this purpose, we define a suitable objective function as follows:

$$\hat{\Theta}(\chi) = \Theta(\chi) + \sum_{i=1}^{m} \rho(a_i, \phi_i(\chi)) + \sum_{i=1}^{n} [\rho(b_i, \varphi_i(\chi)) + \rho(b_i, -\varphi_i(\chi))] \tag{4}$$

Such a function satisfies the inequality constraints of the following equation:

$$(\phi_i(\chi) \geqslant \epsilon, |\varphi_i(\chi)| \leqslant \flat, \flat \rightarrow 0, \epsilon \geqslant \flat, \forall i),$$

which is a model of an optimization problem, where ai and bi are the positive constants; ρ is the penalty function, which modifies the original objective function; ϕi and φi are the constraints. The penalty strategy uses finite difference techniques as a fitness function (FF) in a cloud (see Figure 2).

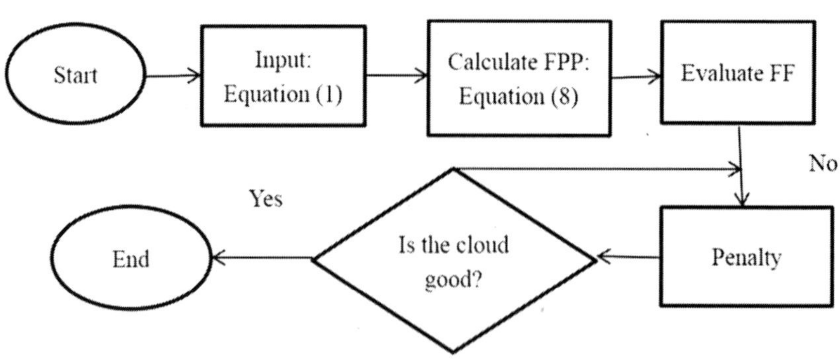


**Figure 2.** The proposed algorithm.

**Fitness function (FF).** The fitness function of a cloud between agents χi and χj is employed to control the process during the system evolution. Equation (1) can be reduced to minimize the problem as follows:

$$\hat{\Xi}_{\Omega}(\hat{\chi}) = \sum_{\chi_i \in \Omega} (\mathcal{D}(\chi_i) - \varphi_i(\chi_i)), \quad (5)$$

where *D(χi):=D(℘)[χi(t)+ϕi(t,χ,u)]*, which satisfies the conditions:

$$\hat{\Xi}_{\partial\Omega}(\chi) = \sum_{\chi_i \in \partial\Omega} (\overline{\mathcal{D}}(\chi_i) - \overline{\varphi}(\chi_i)) = 0, \quad (6)$$

where χ^ is the estimated value of χ. Problems (5) and (6) can be solved by letting the FF:

$$F(\chi) = \hat{\Xi}_{\Omega}(\hat{\chi}) + \varsigma \hat{\Xi}_{\partial\Omega}(\chi), \quad (7)$$

where ς is the penalty parameter.

**Fractional entropy.** In this study, we deal with a measure of entropy imposed by Tsallis [19] with the form:

$$\mathcal{T}_{\gamma}(\phi) = \frac{\int_x [\phi(x)]^{\gamma} dx - 1}{1 - \gamma}, \gamma \neq 1$$

or in the discrete form:

$$\mathcal{T}_{\gamma}(\phi) = \frac{1}{\gamma - 1}(1 - \sum_{k=1}^{m} \phi_i^{\gamma}), \gamma \neq 1$$

Several types of fractional entropy were recommended in [20,21,22,23,24,25]. At this stage, the appropriate amount of information based on observing the appearance of an event has probability *P*. The method is subject to the probability of extinction, which is described by the fractional Poisson process as follows [26]:

$$P_{\wp}(N,u) = \frac{(\sigma u)^N}{N!} \sum_{n=0}^{\infty} \frac{(n+N)!}{n!} \frac{(-\sigma u^{\wp})^n}{\Gamma(\wp(n+N)+1)}$$

where σ∈R is a physical coefficient, ℘∈(0,1], and u is the control in system (1). We let N be the number of agents and J be the average information. The source emits the symbols with probabilities P1,P2,...,PN,respectively, such that Pi = P℘(i,u). Thus, we can calculate the total information by the following equation:

$$\mathcal{I} = \frac{1}{\gamma - 1}(1 - \sum_{i=1}^{m} P_i^{\gamma}), \gamma \neq 1 \tag{8}$$

Note that $0 < \sum_{i=1}^{m} P_i^{\gamma} < 1$. By applying Equation (8) to Equation (7), we conclude the following FFs:

$$\begin{aligned} F(\chi_i) &= \hat{\Xi}_{\Omega}(\hat{\chi}) + \varsigma \hat{\Xi}_{\partial\Omega}(\chi) \\ &\approx \sum_{\chi_i \in \Omega} D^{(\wp)} \chi_i(t) + (\mathcal{I} - \flat) + \varsigma[\sum_{\chi_i \in \partial\Omega} \overline{D}^{(\wp)} \chi_i(t \\ &\quad + (\mathcal{I} - \flat)] \end{aligned} \tag{9}$$

The constraints of Equation (4) are met.

# 5. STABILITY OF THE OUTCOMES

In this section, we discuss the UH stability of Equation (1). We let (X,‖ . ‖) be a Banach space over Rn+ with a maximum norm. We have the following Hyers–Ulam stability (HUS):

## Definition 5.1

We let ϵ be a nonnegative number. Equation (1) is then called stable in the Hyers-Ulam sense if δ>0 exists, such that for every D(χ)∈C(Rn+,X) achieving Equtaion (5):

$$\| \mathcal{D}(\chi) - \varphi(\chi) \| \leqslant \epsilon\beta(\| \chi \|), \epsilon > 0, \tag{10}$$

where β is a positive function; for all χ∈Rn+, an outcome η∈Rn+ exists with the property:

$$\| \chi(t) - \eta(t) \| \leqslant \delta \tag{11}$$

### Theorem 5.1

*We suppose that* D(χ)∈C(R n +,X), *which satisfies Equation (1). If* β *is a linear function in* ‖χ‖,*then every solution is bounded and in HUS.*

### Proof of Theorem 5.1.

Let the condition (10) be achieved. Clearly, we obtain:

$$\begin{aligned}
&\| \chi(t) - \eta(t) \| = \| u(\chi) - \eta(\chi) + \mathcal{D}(\chi) - \varphi(\chi) - \mathcal{D}(\chi) + \varphi(\chi) + \mathcal{D}(\eta) - \varphi(\eta) - \mathcal{D}(\eta) + \varphi(\eta) \| \\
&\leqslant \| \chi \| + \| \eta \| + 2 \| \mathcal{D}(\eta) - \varphi(\eta) \| + 2 \| \mathcal{D}(\chi) - \varphi(\chi) \| \\
&\leqslant \| \chi \| + \| \eta \| + \epsilon\beta(\| \chi \|) + \epsilon\beta(\| \eta \|) \\
&= \| \chi \| + \| \eta \| + \epsilon\beta(\| \chi \| + \| \eta \|) \\
&\leqslant b(1+\epsilon) := \delta,
\end{aligned}$$

where b: = max {‖χ‖+‖η‖,β(‖χ‖+‖η‖)}. Hence, (11) is satisfied and this completes the proof. □.

### Theorem 5.2

*We suppose that* $\mathcal{D}(\chi) \in C(\mathbb{R}^n_+, \mathbb{X})$, *which satisfies Equation (1). We also let* D(χ) *and* φ *be Lipschitz functions with Lipschitz constants* λ1 *and* λ2,*respectively. If contact* 0<λ0<1 *exists, such that:*

$$\lambda := \lambda_0 + \lambda_1 + \lambda_2 < 1, 1 - \lambda_0 \leqslant \epsilon,$$

*then every solution is bounded and HUS.*

**Proof of Theorem 5.2.**

$$\begin{aligned}
&\| \chi(t) - \eta(t) \| = \| \chi(t) - \eta(t) + \mathcal{D}(\chi) - \varphi(\chi) - \mathcal{D}(\chi) + \varphi(\chi) + \mathcal{D}(\eta) - \varphi(\eta) - \mathcal{D}(\eta) + \varphi(\eta) \| \\
&\leqslant \lambda_0 \| \chi - \eta \| + \| \mathcal{D}(\chi) - \varphi(\chi) \| + \| \mathcal{D}(\eta) - \varphi(\eta) \| + \| \varphi(\chi) - \mathcal{D}(\chi) \| + \| \varphi(\eta) - \mathcal{D}(\eta) \| \\
&\leqslant \lambda_0 \| \chi - \eta \| + \epsilon\beta(\| \chi \|) + \epsilon\beta(\| \eta \|) + \lambda_1 \| \chi - \eta \| + \lambda_2 \| \chi - \eta \| + (1 - \lambda_0) \| \chi - \eta \| \\
&\leqslant \lambda_0 \| \chi - \eta \| + \epsilon\beta(\| \chi \|) + \epsilon\beta(\| \eta \|) + \lambda_1 \| \chi - \eta \| + \lambda_2 \| \chi - \eta \| + \epsilon \| \chi - \eta \| \\
&:\leqslant \lambda \| \chi - \eta \| + \epsilon [\beta(\| \chi \|) + \beta(\| \eta \|) + \| \chi \| + \| \eta \|] \\
&\leqslant \lambda \| \chi - \eta \| + \epsilon\overline{\beta},
\end{aligned}$$

where $\overline{\beta} = \max\{\beta(\| \chi \|), \beta(\| \eta \|), \| \chi \| + \| \eta \|\}$. Thus, we conclude that:

$$\| \chi - \eta \| \leqslant \frac{\epsilon\overline{\beta}}{1-\lambda} := \delta$$

Therefore, Eqution (11) is satisfied, and this completes the proof.

**Table 1.** Utility of the fractional cloud entropy system.

| $\wp$ | Time | $P_{\wp}(10)$ | Utility |
|---|---|---|---|
| 1 | 1 | 0.5380 | 0.48 |
| | 2 | 0.5281 | 0.49 |
| | 3 | 0.5139 | 0.491 |
| | 4 | 0.5128 | 0.492 |
| | 5 | 0.5117 | 0.5000 |
| 0.75 | 1 | 0.5800 | 0.5001 |
| | 2 | 0.5740 | 0.5005 |
| | 3 | 0.5585 | 0.5006 |
| | 4 | 0.5573 | 0.5007 |
| | 5 | 0.5561 | 0.5008 |
| 0.5 | 1 | 0.6113 | 0.4000 |
| | 2 | 0.6001 | 0.4001 |
| | 3 | 0.5839 | 0.50011 |
| | 4 | 0.5827 | 0.50012 |
| | 5 | 0.5814 | 0.5002 |

# 6. APPLICATIONS

Table 1 describes the experimental results for N = 10 agents and the different values of the fractional order $\wp$. We suppose the following parameters: $\gamma = 2$ and $\sigma = \varsigma = \rho = 1$, where $\gamma$ is the dimension of the Tsallis entropy, $\sigma$ is the physical coefficient of the Poisson process, $\varsigma$ is the penalty parameter, and $\rho$ is the penalty function that is set as a constant function. The utility of the cloud entropy is evaluated at different times $t \in [1-5]$. The utility seems to increase with increasing time and decreasing the fractional Poisson process ( FPP). The highest number appears at $t = 5$ and $\wp = 0.75$. Increasing the value of $\gamma$ leads to increasing the utility of the cloud (consequently increasing the number of agents in the cloud [16]). The maximum utility goes to the fractional powers $\wp = 0.75$ and 0.5 with increasing time.

# 7. CONCLUSIONS

We have investigated a method for optimizing incomes and the arrangement originated by a system in cloud entropy computing. The method was supplementarily capable, surpassed mathematical software recommendations, and saved the overall cost and the task distribution. However, the objective function was engaged with the concept of the fractional differential equation. By utilizing this equation, we obtained an adaptation to the FF. This procedure was accessible to approximation by employing the fractional Tsallis entropy based on the fractional Poisson process. The stability illuminated the problems on the finite domain, which was suggested in the sense of the UH-stability strategy. The method offered two advantages, namely, it transformed the problem of constrained optimization into an unconstrained one and it had a suitable selection of the fractional order. The technique was a good approximation. Multiple connection could also be proposed by using the aforementioned technique. Furthermore, the method could be increased to a higher dimension when the number of agents in a multi-agent system became large.

# ACKNOWLEDGMENTS

The authors would like to thank the reviewers for their useful comments to improve the paper. This research is supported by Malaysian Ministry of Higher Education under the University Malaya High Impact Research Grant with reference: UM.C/625/1/HIR/MOE/FCSIT/03, and project number: RG312-14AFR from University of Malaya.

## AUTHOR CONTRIBUTIONS

Rabha W. Ibrahim designed the mathematical model; Hamid A. Jalab performed the experiments; Abdullah Gani wrote the paper; all the authors jointly worked on deriving the results. All authors have read and approved the final manuscript.

## REFERENCES

1. Kerrigan, B.; Chen, Y. A Study of Entropy Sources in Cloud Computers: Random Number Generation on Cloud Hosts. In *Computer Network Security*; Springer: Berlin/Heidelberg, Germany, 2012.

2. Tarasov, V.E. *Fractional Dynamics: Applications of Fractional Calculus to Dynamics of Particles, Fields and Media*; Springer: Berlin/Heidelberg, Germany, 2011.

3. Meerschaert, M.M.; Sikorskii, A. *Stochastic Models for Fractional Calculus*; Walter de Gruyter: Berlin, Germany, 2011.

4. Machado, J.T.; Mata, M.E. Pseudo phase plane and fractional calculus modeling of western global economic downturn. *Commun. Nonlinear Sci. Numer. Simul.* **2015**, *22*, 396–406.

5. Machado, J.T.; Mata, M.E. A fractional perspective to the bond graph modelling of world economies.*Nonlinear Dyn.* **2014**, *80*, 1839–1852.

6. Ghosh, S. *Distributed Systems: An Algorithmic Approach*, 2nd ed.; CRC press: Boca Raton, FL, USA, 2014.

7. Ren, W.; Beard, R.W. *Distributed Consensus in Multi-Vehicle Cooperative Control*; Springer: London, UK, 2008.

8. Li, D.; Liu, Q.; Wang, X. Distributed quantized consensus for agents on directed networks. *J. Syst. Sci. Complex.* **2013**, *26*, 489–511.

9. Zeitz, M. Canonical forms for nonlinear systems. In Processdings of IFAC-Symposium, Capri, Italy, 14–16 June 1989; pp. 33–38.

10. Meng, D.; Jia, Y.; Du, J. Nonlinear finite-time bipartite consensus protocol for multi-agent systems associated with signed graphs. *Int. J. Control* **2015**, *88*, 1–24.

11. Lu, C.; Wang, X.; Koutsoukos, X. Feedback utilization control in distributed real-time systems with end-to-end tasks. *IEEE Trans. Parallel Distrib. Syst.* **2005**, *16*, 550–561.

12. Diao, Y.; Gandhi, N.; Hellerstein, J.L.; Parekh, S.; Tilbury, D.M. Using mimo feedback control to enforce policies for interrelated metrics with application to the apache web server. In Proceedings of Network Operations and Management Symposium, 2002 (NOMS 2002), Florence, Italy, 15–19 April 2002; pp. 219–234.

13. Xu, W.; Zhu, X.; Singhal, S.; Wang, Z. Predictive control for dynamic resource allocation in enterprise data centers. In Proceedings of Network Operations and Management Symposium, 2006 (NOMS 2006), Vancouver, BC, Canada, 3–7 April 2006; pp. 115–126.

14. Olfati-Saber, R.; Fax, A.; Murray, R.M. Consensus and cooperation in networked multi-agent systems. *Proc. IEEE* **2007**, *95*, 215–233.

15. Dhage, B.C. Approximation methods in the theory of hybrid differential equations with linear perturbations of second type. *Tamkang J. Math.* **2014**, *45*, 39–61.

16. Ibrahim, R.W.; Jalab, H.A.; Gani, A. Cloud entropy management system involving a fractional power.*Entropy* **2016**, *18*.

17. Dhage, B.; O'Regan, D. A fixed point theorem in banach algebras with applications to functional integral equations. *Funct. Differ. Equ.* **2004**, *7*, 259–267.

18. Pluzhnik, E.; Nikulchev, E.; Payain, S. Concept of feedback in future computing models to cloud systems.**2014**. arXiv:1402.4663.

19. Tsallis, C. *Introduction to Nonextensive Statistical Mechanics*; Springer: New York, NY, USA, 2009.

20. Machado, J.T. Entropy analysis of integer and fractional dynamical systems. *Nonlinear Dyn.* **2010**, *62*, 371–378.

21. Machado, J.T. Entropy analysis of fractional derivatives and their approximation. *J. Appl. Nonlinear Dyn.* **2012**, *1*, 109–112.

22. Machado, J.T. Fractional order generalized information. *Entropy* **2014**, *16*, 2350–2361.

23. Ibrahim, R.W. The fractional differential polynomial neural network for approximation of functions. *Entropy* **2013**, *15*, 4188–4198.

24. Ibrahim, R.W.; Jalab, H.A. Existence of entropy solutions for nonsymmetric fractional systems. *Entropy* **2014**, *16*, 4911–4922.

25. Ibrahim, R.W.; Jalab, H.A. Existence of Ulam stability for iterative fractional differential equations based on fractional entropy. *Entropy* **2015**, *17*, 3172–3181.

26. Laskin, N. Fractional Poisson process. *Commun. Nonlinear Sci. Numer. Simul.* **2003**, *8*, 201–213.

# 5

# CHAPTER

# A MULTI-AGENT SCHOOL SIMULATION BASED ON HIERARCHICAL SOCIAL NETWORKS

**Yu Zhang[1], Jiang Wu[2], Lizhu Ma[3]**

*[1]Department of Computer Science, College of St. Benedict and St. John's University, Collegeville, MN, USA*

*[2]School of Information Management, Wuhan University, Wuhan, China*

*[3]Department of Computer Science, Baylor University, Waco, TX, USA*

## ABSTRACT

The quality of K-12 education has been a very big concern for years. Previous methods studied only one or two factors, such as school choice, or teacher quality, on school performance. There- fore the results they provide can be limited. We propose a multi-agent approach to integrate mul- tiple actors in a school system. These actors include teachers, students, supporting staffs and administrators. The interactions among these actors compose a hierarchical school social network. We first detect the hierarchical community structure in this school network by using an agglo- merative hierarchical algorithm. Existing agglomerative hierarchical algorithms usually calculate similarity or dissimilarity between two clusters by using some measure of distance between pairs of observations. We, however, develop a method that calculates similarity based on social interac- tions between interactions is essential in multi-agent systems. Our algorithm is applied to 15 school districts in Bexar County, Texas, and it provides satisfying results on generating the hie- rarchical structure of all school districts. We then use the detected structure of the social network to evaluate the school system's organization performance. We design and implement a funding evaluation model to decompose the funding policy task into subtasks and then evaluate these subtasks by using funding distribution policies from past years and looking for possible relation- ships between student performances and funding policies. Experiments in the 15 school districts in Bexar County show no significant correlation between student performance and the amount of the funding a school district received.

**Keywords:** Hierarchical Clustering, K-12 Education, Multi-Agent Systems, Social Network Analysis

# 1. INTRODUCTION

Social network analysis has been an emerging field in recent years. It views social relationships in terms of nodes (agents) and edges (ties). Research has shown that social networks play a critical role in determining the way problems are solved, organizations are run, etc. [1]. For a survey on networks, see [2].

A multi-agent system (MAS) is a set of autonomous and interactive entities called agents [3]. Multi-agent system and social network analysis share some similarities (e.g. agents, relationships, etc.). Much research has successfully combined these two together [4] [5]. In multi-agent simulations, when agents communicate with each other or work together on a common goal, agents are often organized into networks.

In this paper, we focus on school networks that contain many actors or agents interacting with each other in school systems. The quality of K-12 education has been a very big concern for years. Many studies have been conducted in the field. There are also many factors that have been studied, such as, school choice [6] [7], school size [8], teacher quality [9] [10], school/school district administrator quality [11] [12], funding [13] [14], etc. Be-cause previous research in this field mostly studied the impact of one or two of those factors on school perfor- mance, the results they provide can be limited.

We propose an approach to integrate multiple factors in a multi-agent school system and show how it can help in school network structure detection and how the detected structure can utilize school policy evaluations. Ac- cording to [15], social networks often have an underlying hierarchical community structure, and finding these community structures is very important for understanding inherent structures for complex social networks [16]. Thus hierarchical clustering algorithms can be useful and appropriate methods to detect multilevel structures of networks [15]. The first step for any hierarchical clustering algorithm is to decide the similarity between two agents or clusters. Because a school system consists of many interacting agents, for example, students and stu- dents, students and teachers, regular education teachers and special education teachers, teachers and administra- tion staff, etc., it is regarded as an interaction-based system. Therefore we use the interaction level between two agents or clusters as the similarity measure. Once the hierarchical structure of the school system is detected, it can be used to evaluate policies that may affect school performance.

The purpose of this work is to study structures of social networks with a focus on social interactions by de- tecting underlying hierarchical structures. The detected structure of social networks is then used to help evaluate their organization performance. There are mainly two steps. The first step is to detect hierarchical community structure in school networks by using an agglomerative hierarchical algorithm. Agglomerative or bottom-up al- gorithms start with each node in its own singleton cluster, and at each step merge these clusters into larger ones until all clusters are merged into one big cluster [9]. Existing agglomerative hierarchical algorithms usually cal- culate similarity or dissimilarity between two clusters by using some measure of distance between pairs of ob- servations. We, however, develop a method that calculating similarity based on social interactions between in- teractions is essential in multi-agent systems. Our algorithm is applied to 15 school districts in Bexar County, Texas, and it provides satisfying results on generating the hierarchical structure of all school districts.

The second goal is to study multiple factors that affect organization performance with the help of organization structure. In the school system network, how funding is distributed in the school system is a very important matter; however, research in the field has not been able to draw any sound conclusions yet. We design and im- plement a funding evaluation model, which is based on the MAXQ algorithm [17], to decompose the funding policy task into subtasks and then evaluate these subtasks by using funding distribution policies from past years and looking for possible relationships between student performances and funding policies. We study funding policies for 15 Bexar county school districts in three years and evaluate these policies. This model first decom- poses the whole funding distribution task into several subtasks. Then, for each school district, we evaluate its funding policy from 2008-2011 by evaluating its performance in these three years. We conclude that increasing funding does not necessarily have a positive impact on student performance and it is generally not the case that the more is spent, the better.

# 2. SCHOOL SYSTEM EVALUATION

Different actors/agents in a school system may play different roles and have different impact on school perfor- mance. In the followings, we review the literatures that discuss the roles of superintendent, principal and teacher and how school quality and student performance might be affected by them.

## 2.1. Funding

School funding is an important and highly contested matter surrounding education. There is conflicting evidence on whether or not an increase in school funding will truly produce an increase in student achievement or wheth-er it is not the amount of funding but how funding is spent that is truly important.

Various research papers and programs provide general data on school funding. Reference [18] indicated that half of all property taxes go to funding public schools. Despite concerns that this figure is too high, it has proven difficult to lower property taxes and improve school performance. The 2008 PBS program Where We Stand: America's Schools in the 21st Century indicated that 10% of school funding comes from federal sources, with the majority coming from state and local sources.

The general consensus is that schools are underfunded and require more funding in order to function properly. Reference [14] indicated this when he stated that investment in the human, social, and physical capital of a school can have a strong positive impact on student achievement. However, [13] indicated that the effects of in- creasing per-pupil spending on student achievement are not necessarily positive. Although school funding has been steadily increasing since 1970, American schools are still falling behind schools in other countries. There are a number of concerns regarding these figures. For one, although funding has increased overall, there are concerns that there is a large funding gap between schools, with some schools receiving insufficient funding and other schools receiving too much funding. Also, there are concerns that money is not being spent in ways that have a significant impact on student achievement.

Overall, schools require funding to be effective in increasing student achievement, but too much funding produces diminishing returns. Schools must therefore be able to spend money intelligently. Anderson recommended a targeted approach, identifying the schools and programs that must be invested in to increase student achievement and then applying funding to those areas to achieve the desired result.

## 2.2. Superintendent

Report from [12] found that superintendent quality does affect school performance. They believed that superin- tendent can affect school quality in a variety of ways (for example, recruiting, training, and rewarding high-quality principals and occasionally teachers). They also suggested that the quality of a superintendent can be measured as the additional salary paid to school superintendents over and above the normal determinants of sal- ary. The McREL (Mid-continent Research for Eudcation and Learning) research team found a statistically sig-nificant relationship between district leadership and student achievement [19]. They also identified five district responsibilities that are significant to student achievement, such as non-negotiable goals for achievement and in- struction, board alignment and support of district goals, etc.

## 2.3. Principal

Reference [11] found a positive relationship between principal experience and school performance, but they found little evidence of any relationship between school performance and principal education. Reference [20] study several types of principal leadership. They suggested that transformational and shared instructional lea- dership coexisting in an integrated form of leadership is the best for improving school performance.

## 2.4. Teacher

Teachers are directly involved with students in everyday school life. It is very obvious that they have a substan- tial impact on student achievement. Teacher's ability is the key to student achievement, as discussed in [21]. Reference [22] argued that only true way for schools to improve student achievement is through knowledgeable and skilled teachers. Even if schools have large amounts of funding, they may still have low student achievement because they don't invest in high-quality teachers and instead devote money to "fad" programs which don't have any effect on student achievement.

Exactly what makes a high-quality teacher is a topic that is still up for debate. For instance, the most common measures used to gauge a teacher quality and in determining a teacher's salary are the teacher's education and experience. Reference [23] argued that teachers who have advanced degrees in their field produce higher achievement gains in students than teachers who do not. However, [10] [24] reported that there is no evidence or common consensus that teachers with high scholastic aptitude really do produce significant achievement gains in students. But they do believe that that some forms of training can help improve teacher ability. Reference [21] reported that teacher experience produces significant gains in student test scores. Hanushek, however, stated that the connection between teacher experience and student achievement is actually very weak.

Teacher salary is another area that could potentially have a positive impact on student achievement. Reference [25] found no clear link between teacher performance pay and student achievement, although he remarked that an association between these two factors may appear in countries where teacher salaries are relatively low. Ref- erence [24], however, stated that teacher salary may be an indicator of student achievement, but only if it is linked to teacher qualities that are associated with improved student achievement. For instance, salary may not be linked to achievement if it is determined by education and experience, but a connection may appear if salary is determined by student test score gains.

## 2.5. Student

A very important, if not the most important goal of schools is to increase the academic achievement of students. When "achievement" is mentioned it is generally implied to mean the performance of a student in school, measured using the student's scores on standardized tests and/or the student's educational growth over time.

Researchers believed that there are factors and policies that may affect student performance. References [26] [27] indicated that the socioeconomic background of a student plays a major role in his/her achievement. Fami- lies from high socioeconomic strata tend to be more involved in the educational system and take a more active role in helping the student perform better.

The people that students interact with, and not just their teachers, can affect their achievement. Reference [27] indicated that students with high educational ability can increase not just their own achievement but the achievement of their peers. Reference [28] indicated that student achievement can also be improved by strong support from the student's parents and the community he/she lives in. Even students who are at risk for low achievement can show significant improvement with strong parental support.

The effects of specific educational policies on student achievement have also been studied. Reference [29] examined the effects of public monitoring of public schools on student achievement. Because the effort that a school makes in improving student achievement is not perfectly observable by parents, schools have an incen- tive to under-provide effort. [29] reported that better and more direct monitoring of public schools will cause them to provide more educational effort, which in turn increases student achievement. Reference [30] examined the effect of high school graduation requirements on student achievement and found that few students are affected by them. Most students tend to either take more courses than are required or else take courses that don't affect their achievement. School bureaucracy was studied by [26] , who concluded that the relationships between bureaucracy and student performance is negative.

# 3. THE MODEL

## 3.1. Overall Algorithm

The overall system run, see Algorithm 1, is that it first takes the graph G, policy such as the funding policy, and the reward for the policy, as the input. First, the similarity matrix is generated (introduced in Section 3.2). Then by calling the hierarchical clustering algorithm (also introduced in Section 3.2), the hierarchical structure of the network is detected. Finally, the funding evaluation algorithm is called (introduced in Section 3.3) to decompose funding distribution task into subtasks and study policies for each subtask by calculating the reward for each subtask.

**Algorithm 1.** Hierarchical structure detection and analysis.

```
hierarchical_lustering(s_m)
    S0 = φ
    initialize each node a, b, … in its own cluster ai, bi, …
    Do until Si is the only cluster left
        merge two clusters, say ai and bi with highest similarity d, into ni
        (Si\{ai, bi}) ∪ ni//add ni to Si and delete ai and bi
     return Si
```

## 3.2. Interaction-Based Similarity Measure Algorithm

Suppose there is a set of S of N nodes. The input to the clustering algorithm can be defined as a similarity matrix. If the set S has N nodes, there should be $\binom{N}{2}$ pairwise similarities.

Definition 1 Similarity Matrix. A d × d similarity matrix D on a set S is $d(a,b)$ where $a,b \in S$. $d(a,b)$ is a quantitative measure of the similarity between two nodes a and b . $d(a,b)$ is both reflexive and symmetric, i.e. we have $d(a,a)=0$ and $d(a,b)=d(b,a)$ for all $a,b \in S$.

The network is essentially built up by a multi-agent system, where agents interact with each other and these in- teractions have different levels because some agents interact more closely while others have more loose interac- tions. The first step of the interaction-based similarity measure algorithm is to calculate similarity between each pair of nodes. This similarity measure depends on the specific problem. For instance, in biological studies, it could be real physical distance between two nodes. In film actors' collaboration networks, the similarity could be calculated by how many film actors have appeared in together [31].

In interaction-based social networks, it would be different because there is no real physical distance measure in the system. Therefore we propose a way to define similarity (a, b). Because nodes in our system are agents and agents have interactions with each other, we use their interaction level to be the similarity measure. Agents (nodes) that share more frequent interaction have higher similarity while agents that share less frequent interaction have lower similarity. For instance, in our school system, student-teacher pair should have higher similarity than stu- dent-principal pair because students interact more often with their teachers than that of the principle.

Definition 2 Social Interaction [32] . Social interaction between two agents a and b occurs usually either when 1) a initiates it by performing a social action intended to influence b, then b "reacts", i.e., performs an action in consequence of a's action, or when 2) each of them independently starts a social action bearing upon the other, and each reacts to the other's action.

Definition 3 Similarity Measure. The similarity $^{d(a,b)}$ between nodes a and b is the interaction level be- tween a and b.

The interaction level is essentially domain dependent. For example, in our school systems, the interaction level is evaluated based on the actual interaction time length between students and teachers, students and other staffs, students and the principle, etc.

The output of a hierarchical clustering procedure is a dendrogram.

Definition 4 Dendrogram [33]. Given a finite set $S_0$ (initial set) with cardinality $^{N=|S_0|}$, a dendrogram is a list of N – 1 triples ($a_i$, $b_i$, $\delta_i$), i = 0, N ? 2, such that $\delta_i$ Î [0, ∞) is the distance between $a_i$ and $b_i$ and $a_i$, $b_i$ Î $S_i$, where $S_{i+1}$ is recursively defined as ($S_i$\{$a_i$, $b_i$}) ∪ $n_i$, and $n_i$ Ï ($S_i$\{$a_i$, $b_i$}) is a label for a new cluster.

The set $S_0$ are the initial data points. In each step, $n_i$ is the new cluster which is formed by joining the clusters $a_i$ and $b_i$ at the distance $\delta_i$. The order of the clusters within each pair ($a_i$, $b_i$) does not matter. "\" represents relative complement. $S_i$\{$a_i$, $b_i$} is the set of elements in $S_i$ but not in {$a_i$, $b_i$}. The procedure contains N ? 1 steps, so that the final state is a single cluster which contains all N initial nodes.

The dengrogram represents a recursive procedure where at each step a new cluster $n_i$ is formed from two clus- ters $a_i$ and $b_i$ based on their distance $\delta_i$. In every step, a new cluster is added to the set and the previous two indi- vidual nodes that formed the cluster are eliminated from the set. See Algorithm 2.

**Algorithm 2**. Hierarchical clustering algorithm.

---

*hierarchical_lustering*(s_m)

$S_0 = \phi$

initialize each node a, b, … in its own cluster $a_i$, $b_i$, …

Do until $S_i$ is the only cluster left

merge two clusters, say $a_i$ and $b_i$ with highest similarity d, into $n_i$

$(S_i \setminus \{a_i, b_i\}) \cup n_i$//add $n_i$ to $S_i$ and delete $a_i$ and $b_i$

return $S_i$

---

After the step where nodes are clustered into their first cluster, which uses node similarity d, we deal with clus- ters directly. We use average linkage, which is the average similarity between all pairs of nodes in the two clus- ters, to calculate cluster similarity:

$$\delta(a_i, b_i) = \frac{\sum a \in a_i, \sum b \in b_i, d(a, b)}{|a_i| \times |b_i|} \tag{1}$$

where $a_i$ and $b_i$ are the cluster pair and a and b are nodes within them.

## 3.3. Funding Evaluation

Some of the problems of reinforcement learning tend to be very large in scale. Thus the hierarchical approach to reinforcement learning has been developed to decompose complex problems by using temporal abstraction and hierarchical control [34]. Among hierarchical reinforcement learning algorithms, the MAXQ approach is consi- dered one of the most effective methods [35]. The MAXQ approach starts with a decomposition of a core MDP into a set of subtasks. All the tasks form a task graph hierarchically. As the task graph decomposes the action space of the problem, the MAXQ graph decomposes the value function of the problem [8].

We build a funding policy evaluation system by using the MAXQ decomposition. Funding is usually distri- buted into several areas such as teachers, staffs, administrators, regular education students, special education stu- dents, gifted education students, ESL students (English as Second Language), and career education students. Therefore, the model breaks the major task into several subtasks. Then each subtask can be accomplished by one or more primitive actions. Subtasks could be different funding distribution areas, which can be accomplished by different primitive actions (i.e. funding distribution methods). A reward that is evaluated by student performance is returned to each finished action. Then once different funding policies are inputted into the model, it will eva- luate them, compare the rewards, and finally return the one with highest reward.

Definition 5 MAXQ Task. The overall MAXQ task is to find an optimal funding policy M, which is defined over a set of states S (i.e. the funding amount received by each type of the agents such as teacher, student, and administrator) and actions A (i.e. giving an amount of funding to each type of the agents) with reward function $R(s'|s,a)$ (the reward received upon entering a new state $s'$ after receiving a funding amount in the state s) and transition probability function $R(s'|s,a)$(the probability of entering state $s'$ as a result of receiving a funding amount in s).

The basic idea of the MAXQ approach is that it decomposes the whole task into a set of subtasks, which may further be decomposed into smaller subtasks [35] . This structure forms a hierarchy tree whose leaves are primi- tive actions. Primitive actions return the rewards for performing the actions. Each subtask has some termination conditions, which are the conditions that once fulfilled the control of program returns to the parent subtask.

We say that hierarchical policy for a MAXQ graph is a set of funding policies, $\pi = (\pi_1, \pi_2, \cdots, \pi_n)$ one for each subtask. As the action space of the problem is decomposed by the task graph, we decompose the action-value function $Q^{\pi}(p,s,a)$, i.e. the total expected reward of performing action a in subtask p and then follow- ing the hierarchical policy $\pi = (\pi_1, \pi_2, \cdots, \pi_n)$ into two components, see Equation (2).

The first component $V^{\pi}(a,s)$, see Equation (3), is the expected total reward received while executing action a in state following policy $\pi$, and the second componentin Equation (4), which is called as the completion function, $C^{\pi}(p,s,a)$ is the expected total reward of completing parent task $p$ following policy $\pi$ after a has returned. Thus we have the MAXQ decomposition equations:

$$Q^{\pi}(p,s,a) = V^{\pi}(a,s) + C^{\pi}(p,s,a) \tag{2}$$

where

$$V^{\pi}(a,s) = \begin{cases} Q^{\pi}(a,s,\pi_i(s')) & \text{if } i \text{ is composite subtask} \\ \sum_{s'} P(s'|s,i)R(s'|s,i) & \text{if } i \text{ is a primitive action} \end{cases} \tag{3}$$

and

$$C^{\pi}(p,s,a) = \sum_{s',n} P(s',n|s,a)\gamma^{n} Q^{\pi}(p,s',\pi(s')) \tag{4}$$

Note: n here stands for time steps and $\gamma$ is the time discount factor which determines the importance of fu- ture rewards. $s'$ is the new state that is entered after performing action a instate s.

# 4. EXPERIMENT

We implement a school system network by using real school data provided by the Texas Education Agency[1]. Bexar County in Texas is the geographical area we focus on. There are 15 public school districts in Bexar as of 2011 (see Table 1). Bexar County is used because it consists of mixed type school districts. In 2011, among the 15 school districts, 11 were rated as "academically acceptable" and 4 were rated as "recognized".

There are many factors that may affect school performance. Previous research in this field mostly studied the impact of one or two of those factors. We observe the emergence of school system performance based on a com- prehensive list of agents/factors and the interactions between them.

School funding is a very important matter surrounding education. There is conflicting evidence on whether or not an increase in school funding will truly produce an increase in student achievement or whether it is not the amount of funding but how funding is spent that is truly important [13] [14] [18] . We design a model that ana- lyzes funding distribution records and their relations to student performance in the past year based on a hierar- chical decomposition.

## 4.1. School System Structure

Most of the school districts in the United States are composed of public schools, private schools, and charter schools. For each school district, there is a school board, which appoints a superintendent. Each school has its own principal and teachers.

We design a multi-agent system to model a school district system. Our system includes many actors/agents such as students, teachers, principals, superintendent, etc. Our method is different from existing ones because ours observes the emergence of school system performance based on a comprehensive list of agents and the interac- tions between them. An abstract representation of a single school is presented Figure 1.

Each node in the graph represents one type of agent. Edges represent interaction between nodes. Students are the largest group in the school system. They interact with each other. They also learn from the teachers. Teachers teach students and report to school administrative staffs (e.g. principal). School administrative staffs report to central administrative staffs (e.g. superintendent).

The simulation grounds on individual agents. The actual size of our system is presented in previousTable 1. To give readers a sense of complexity, Figure 1 visualizes a graph of a small sample school district, where there are only 10 students, 2 teachers, 1 other staff, 1 central administrative staff, and 1 school administrative staff. The graph is generated by a free and open-source application called NodeXL[2].

## 4.2. Experiment Settings I: Clustering

The experiment is composed of two parts: hierarchical structure detection and funding policy evaluation. In the first part, we use a hierarchical clustering algorithm to detect hierarchical structure in the network. Then, in the second part, we use the structure detected to help to evaluate previous funding policies.

As shown in Figure 2, the system has many agents connecting with each other. However, the structure of the network cannot be seen directly from the graph visualization, so it needs to be found. Therefore, here we use the agglomerative hierarchical clustering algorithm to detect underlying hierarchical community structure in the school system.

Students and teachers are further divided into 10 groups: regular education students, special education students, ESL (English as Second Language) education students, career education students (CES), gifted education students (GES), regular education teachers (RET), special education teachers (SET), ESL education teachers (EET), career education teachers (CET), and gifted education teachers (GET). Table 2(a) and Table 2(b) show the number of agents in each group.

**Table 1**. School district profiles 2011.

| District name | # of Schools | Total Students | Total Teachers | Total Central Admin | Total School Admin | Total Other Staffs | Total Agents |
|---|---|---|---|---|---|---|---|
| Alamo Heights | 6 | 4744 | 335 | 6.04 | 12.08 | 253.68 | 5350.8 |
| Harlandale | 30 | 14,846 | 970 | 20.86 | 62.58 | 1022.14 | 16921.58 |
| Edgewood | 22 | 11,904 | 700 | 0 | 34.82 | 905.32 | 13632.14 |
| Randolph Field | 3 | 1167 | 86 | 5.04 | 3.36 | 72.24 | 1333.64 |
| San Antonio | 99 | 54,894 | 3437 | 76.31 | 152.62 | 3968.12 | 62528.05 |
| South San Antonio | 18 | 9860 | 687 | 29.36 | 44.04 | 719.32 | 11339.72 |
| Somerset | 7 | 3779 | 258 | 5.89 | 17.67 | 318.06 | 4378.62 |
| North East | 73 | 66,364 | 4377 | 0 | 171.14 | 3936.22 | 74848.36 |
| East Central | 15 | 9571 | 573 | 12.06 | 24.12 | 603 | 10783.18 |
| Southwest | 14 | 11,779 | 735 | 15.02 | 45.06 | 720.96 | 13295.04 |
| Lackland | 2 | 985 | 80 | 1.68 | 3.36 | 80.64 | 1150.68 |
| Ft. Sam Houston | 2 | 1427 | 118 | 2.31 | 6.93 | 103.95 | 1658.19 |
| Northside | 107 | 94,632 | 6117 | 0 | 255.5 | 6387.5 | 107392 |
| Judson | 30 | 22,016 | 1437 | 0 | 89.37 | 1459.71 | 25002.08 |
| Southside | 9 | 5310 | 375 | 7.49 | 22.47 | 352.03 | 6066.99 |

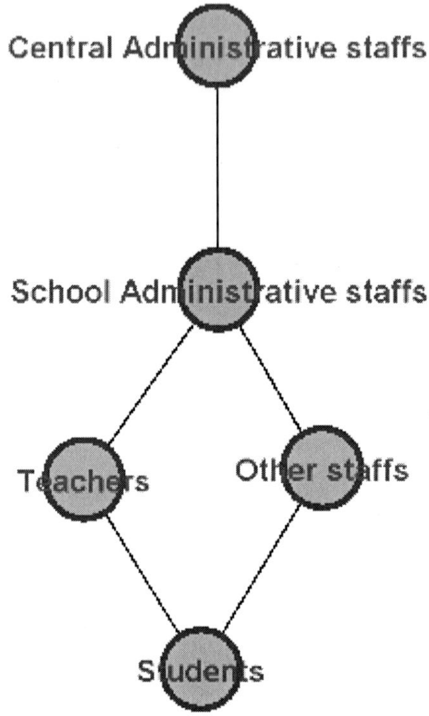


**Figure 1**. School organization. Note: This graph only presents one school as an example of the organization because of space limit in this paper. Our datasets contains 15 school districts and in each district, there are 10 - 20 schools that are all connec- ted to the central administrative staff in the graph.

Before we start, similarity between each pair of nodes must be calculated. Because nodes in the system are agents, and agents have interactions with each other, the similarity measure is set to be their interaction levels. The higher the interaction level, the higher the similarity. We define the interaction level as a function of the time length of agents' interactions, as shown in Equation (5):

$$interaction\_level = f(time\_length) \quad (5)$$

The more time the two agents spend on interaction, the higher the interaction level.

Central and school administrative staffs generally do not interact with students directly very often, but they can still have an effect on student performance and school quality [11] [12]. However, they should work with each other and teachers more directly (through recruiting, training, and rewarding high-quality principals and teachers) [12]. All students should receive regular education from regular education teachers, so they interact with each other frequently. There are also 4 other kinds of education: special, ESL, gifted, and career education. Students who also receive these four kinds of education interact with teachers of these four kinds directly. Although these students spend less time with them than with regular teachers, in order to distinguish these students from regular students, we set these interaction levels higher than the interaction levels of these students with regular teachers.

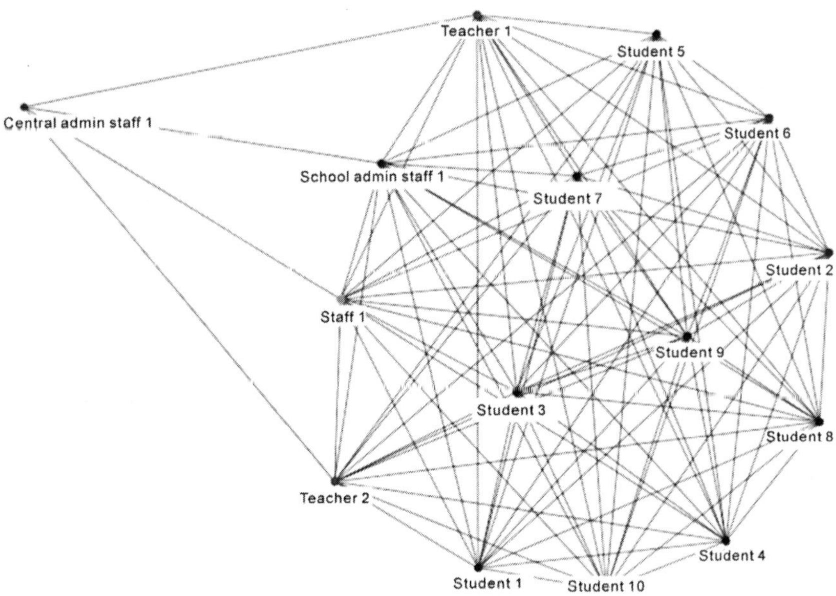


**Figure 2**. School district agents and interactions figure.

Table 3(a) shows interaction levels between all kinds of agents except interactions within and between students and teachers, which are specified by groups in Table 3(b). Interaction levels are determined by the actual interac- tion time lengths between agents, such as students and teachers, students and staffs, etc. We normalize the actual time lengths to a scale of 0 - 9. 0 means the least interaction, while 9 means the highest interaction.

The above tables are the input matrix for the clustering algorithm. The output is a dendrogram that shows out the underlying structure. We write the simulation code in Python and use the fast cluster library and its interface to Python to accomplish the clustering process because it has proven to perform well in terms of complexity[3]. Because fast cluster uses dissimilarity rather than similarity between nodes as the input, the matrix for the simu- lation program is calculated by 9 - d where d is the similarity measure presented in Table 3(a) and Table 3(b).

## 4.3. Experiment Settings II: Funding Policy Evaluation

The second setting part of the experiment focuses on evaluating funding policy. In our experiment, the task (i.e. the funding) is decomposed into several subtasks, which are the regular education area, special education area, bi- lingual education area, career and technical education area, and gifted education area, as shown in Figure 3.

The "primitive actions" are the actual funding distributions shown in Table 4. We collect the data for consecu- tive three years from 2008-2010. Table 3 shows the 2010 data.

After each subtask chooses an action, the reward for that action is returned. Each subtask will iterate over all actions and rewards will be compared in the end, and the action that returns highest reward will be chosen. The reward that is returned for each action is a function of student performance.

$$\text{reward} = f\left(\text{student\_performance}\right) \tag{6}$$

where student_ performance is evaluated by the TAKS (Texas Assessment of Knowledge and Skills) test passing rates each year (also see Table 4).

**Table 2.** (a) Student groups data; (b) Teacher groups data.

(a)

| District Name | Total Students | Special Student | ESL Student | Career Student | Gifted Student | Regular Student |
|---|---|---|---|---|---|---|
| Alamo Heights | 4744 | 284.64 | 237.2 | 284.64 | 759.04 | 3178.48 |
| Harlandale | 14,846 | 1336.14 | 2375.36 | 3711.5 | 890.76 | 6532.24 |
| Edgewood | 11,904 | 1190.4 | 2261.76 | 2261.76 | 952.32 | 5237.76 |
| Randolph Field | 1167 | 81.69 | 0 | 93.36 | 93.36 | 898.59 |
| San Antonio | 54,894 | 5489.4 | 8783.04 | 10978.8 | 3293.64 | 26349.12 |
| South San Antonio | 9860 | 788.8 | 1479 | 1972 | 394.4 | 5225.8 |
| Somerset | 3779 | 340.11 | 377.9 | 718.01 | 151.16 | 2191.82 |
| North East | 66,364 | 6636.4 | 5972.76 | 9954.6 | 4645.48 | 39154.76 |
| East Central | 9571 | 1052.81 | 861.39 | 1627.07 | 574.26 | 5455.47 |
| Southwest | 11,779 | 1295.69 | 1413.48 | 2473.59 | 471.16 | 6125.08 |
| Lackland | 985 | 98.5 | 19.7 | 157.6 | 49.25 | 659.95 |
| Ft. Sam Houston | 1427 | 171.24 | 57.08 | 199.78 | 85.62 | 913.28 |
| Northside | 94,632 | 11355.84 | 6624.24 | 17033.76 | 9463.2 | 50154.96 |
| Judson | 22,016 | 2201.6 | 1541.12 | 5504 | 1761.28 | 11008 |
| Southside | 5310 | 531 | 637.2 | 318.6 | 318.6 | 2548.8 |

(b)

| District Name | Total Teachers | Regular Teacher | Special Teacher | ESL Teacher | Career Teacher | Gifted Teacher |
|---|---|---|---|---|---|---|
| Alamo Heights | 335 | 234.5 | 30.15 | 16.75 | 3.35 | 33.5 |
| Harlandale | 970 | 727.5 | 116.4 | 58.2 | 38.8 | 29.1 |
| Edgewood | 788 | 543.72 | 94.56 | 94.56 | 23.64 | 23.64 |
| Randolph Field | 86 | 72.24 | 6.02 | 0 | 0.86 | 4.3 |
| San Antonio | 3437 | 2199.68 | 378.07 | 343.7 | 137.48 | 240.59 |
| South San Antonio | 687 | 377.85 | 68.7 | 116.79 | 27.48 | 68.7 |
| Somerset | 258 | 196.08 | 25.8 | 20.64 | 10.32 | 5.16 |
| North East | 4377 | 3107.67 | 437.7 | 218.85 | 131.31 | 393.93 |
| East Central | 573 | 338.07 | 63.03 | 40.11 | 17.19 | 40.11 |
| Southwest | 735 | 536.55 | 73.5 | 73.5 | 29.4 | 22.05 |
| Lackland | 80 | 61.6 | 6.4 | 0.8 | 3.2 | 0.8 |
| Ft. Sam Houston | 118 | 92.04 | 12.98 | 2.36 | 2.36 | 5.9 |
| Northside | 6117 | 3914.88 | 795.21 | 489.36 | 183.51 | 428.19 |
| Judson | 1437 | 1120.86 | 129.33 | 28.74 | 71.85 | 57.48 |
| Southside | 375 | 258.75 | 41.25 | 18.75 | 15 | 18.75 |

## 4.4. Results

### 4.4.1. Hierarchical Clustering

The interaction-based clustering algorithm works for all 15 school districts. As an example, the output dendro- gram for Lackland school district is shown in Figure 4. We choose this school district to present because it is one of the best-performing school districts in Bexar County. The dendrogram is generated by Python's matplot- lib library. Students and teachers are first clustered into each subgroup. Then it is clustered together with other staffs (these are shown on the bottom of Figure 4). School and school district staffs are clustered together as shown on the left. Then, they are clustered into one final cluster, as shown on top.

The number on the y axis shows the distance of the two clusters that are formed together.

### 4.4.2. Funding Evaluation

Figure 5 represents the relations between the TAKS passing rates and expenditure per student of all three years. We observe mixed-up correlations between two data. For example, for some school districts, when expenditure per pupil increases, student performance decreases (e.g. Northside and Southwest), or when expenditure per pupil decreases, student performance increases (e.g. Alamo Heights and Ft Sam Houston), or students perform better while their expenditures increase (e.g. Somerset and Southside), or students do not performance significantly dif- ferently no matter if the funding is increased or decreased (e.g. Randolph Field and Lackland). Therefore, we conclude that it is not necessarily the case that the more spend the better. The data on Bexar County do not show a consistent relation between changes in funding and student performance.

**Table 3**. Interaction levels a & b.

(a)

| | Student | Teacher | Other staff | School admin staff | Central admin staff |
|---|---|---|---|---|---|
| Student | / | / | 5 | 2 | 0 |
| Teacher | / | / | 5 | 4 | 2 |
| Other staff | 5 | 5 | 9 | 4 | 2 |
| School admin staff | 2 | 4 | 4 | 9 | 8 |
| Central admin staff | 0 | 2 | 2 | 8 | 9 |

(b)

| | RES | SES | EES | GES | CES | RET | SET | EET | GET | CET |
|---|---|---|---|---|---|---|---|---|---|---|
| Regular education student (RES) | 9 | 7 | 7 | 7 | 7 | 9 | 6 | 6 | 6 | 6 |
| Special education student (SES) | 7 | 9 | 7 | 7 | 7 | 8 | 9 | 6 | 6 | 6 |
| ESL education student (EES) | 7 | 7 | 9 | 7 | 7 | 8 | 6 | 9 | 6 | 6 |
| Gifted education student (GES) | 7 | 7 | 7 | 9 | 9 | 8 | 6 | 6 | 9 | 6 |
| Career education student (CES) | 7 | 7 | 7 | 7 | 9 | 8 | 6 | 6 | 6 | 9 |
| Regular education teacher (RET) | 9 | 8 | 8 | 8 | 8 | 9 | 7 | 7 | 7 | 7 |
| Special education teacher (SET) | 6 | 9 | 6 | 6 | 6 | 7 | 9 | 7 | 7 | 7 |
| ESL education teacher (EET) | 6 | 6 | 9 | 6 | 6 | 7 | 7 | 9 | 7 | 7 |
| Gifted education teacher (GET) | 6 | 6 | 6 | 9 | 6 | 7 | 7 | 7 | 9 | 7 |
| Career education teacher (CET) | 6 | 6 | 6 | 6 | 9 | 7 | 7 | 7 | 7 | 9 |

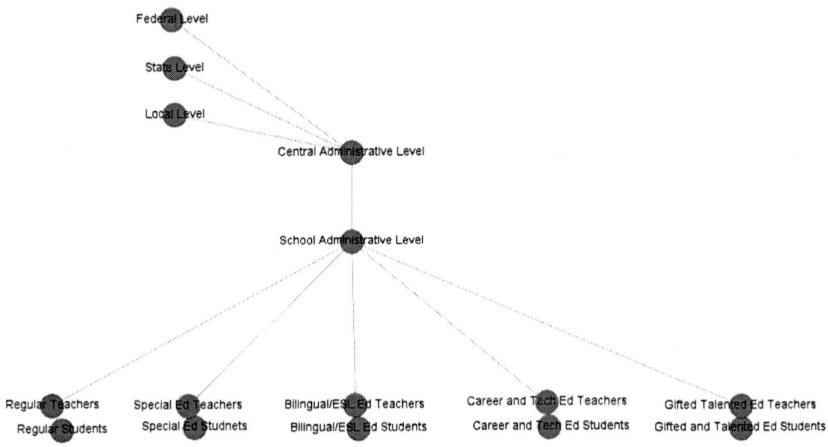


**Figure 3**. Funding decomposition.

**Table 4**. Funding data and TASK passing rates for the year of 2010.

| District Name | TAKS | Total Expenditure | Regular Education | Special Education | ESL Education | Career Education | Gifted Education |
|---|---|---|---|---|---|---|---|
| Alamo Heights | 87 | 43,715,556 | 29726578.08 | 6994488.08 | 0 | 437155.56 | 874311.12 |
| Harlandale | 71 | 127,324,400 | 68,755,176 | 16,552,172 | 8,912,708 | 3,819,732 | 0 |
| Edgewood | 59 | 111,248,406 | 50061783.7 | 13349808.72 | 7787388.12 | 3337452.18 | 0 |
| Randolph Field | 91 | 12,701,043 | 9525782.25 | 1397114.73 | 0 | 127010.43 | 508041.72 |
| San Antonio | 62 | 504,207,177 | 226893229.7 | 90757291.86 | 40336574.16 | 15126215.31 | 5042071.11 |
| South San Antonio | 63 | 93,898,291 | 41315248.04 | 14084743.65 | 11267794.92 | 3755931.64 | 1877965.82 |
| Somerset | 65 | 33,487,386 | 15739071.42 | 4688234.04 | 1674369.3 | 1339495.44 | 669747.72 |
| North East | 81 | 559,721,134 | 347027103.1 | 123138649.5 | 5597211.34 | 10791034.02 | 5597211.34 |
| East Central | 70 | 74,311,079 | 4161204.24 | 14862215.8 | 2229332.37 | 2229332.37 | 1486221.58 |
| Southwest | 67 | 99,087,846 | 59452707.6 | 14863176.9 | 990878.46 | 3963513.84 | 0 |
| Lackland | 87 | 11,806,204 | 7674032.6 | 2361240.8 | 118062.04 | 236124.08 | 118062.04 |
| Ft. Sam Houston | 75 | 19,322,253 | 11013684.21 | 6376343.49 | 0 | 386445.06 | 193222.53 |
| Northside | 80 | 761,630,471 | 472,210,892 | 167558703.6 | 0 | 22848914.13 | 7616304.71 |
| Judson | 69 | 174,658,801 | 104795280.6 | 33185172.19 | 5239764.03 | 6986352.04 | 0 |
| Southside | 68 | 46,849,510 | 26704220.3 | 6090436.3 | 2810970.6 | 1873980.4 | 0 |

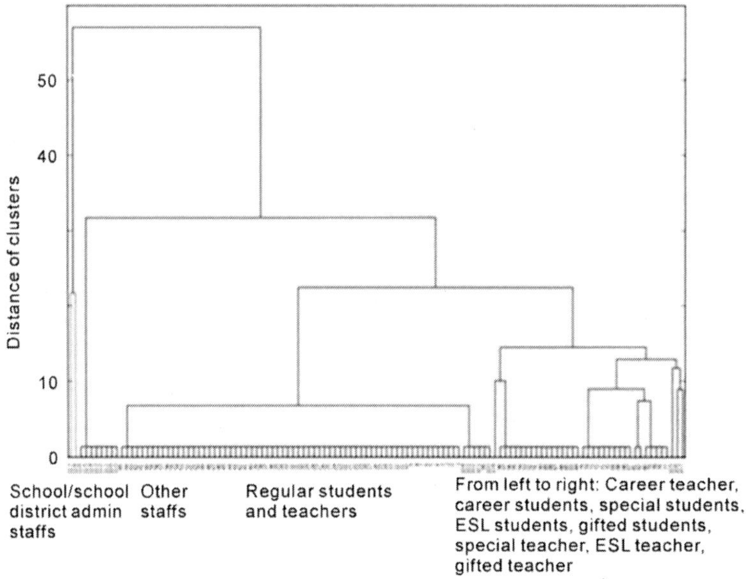


**Figure 4**. The dendrogram for Lackland.

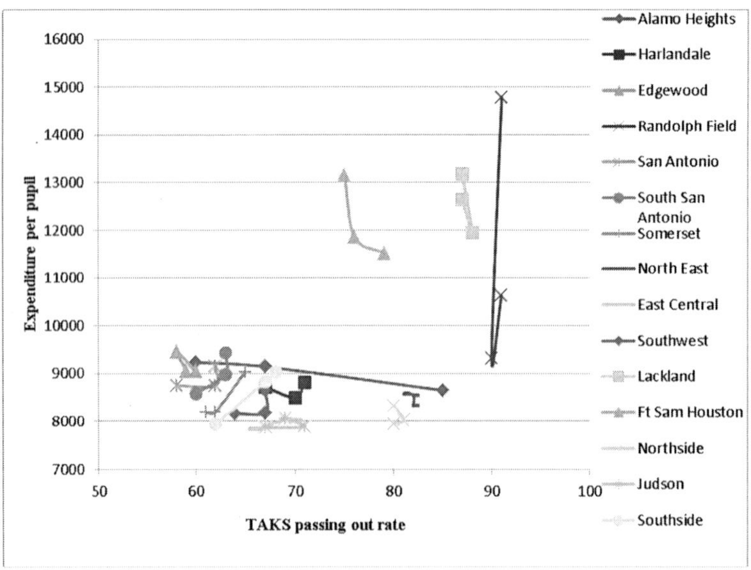


**Figure 5**. Relations between the TAKS passing rates and expenditure per student of all three years. Note: the three points on each line represents the three years: Left point—2008, middle point—2009, right point—2010.

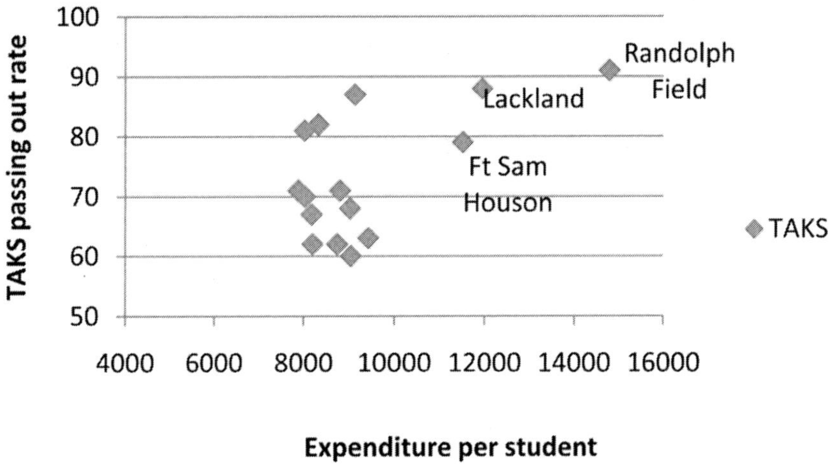


**Figure 6**. Relations between best TAKS passing rate among three consecutive years and ex- penditure per student.

Figure 6 shows our results of comparing all school districts together with their most effective-policy year. For school districts that have their total expenditure per student under 10,000, more funding does not seem to make them perform better than other schools. These districts are shown without labels in the figure. However, Ft Sam Houston, Lackland and Randolph Field districts (shown in the figure with labels) have relatively high expenditure per pupil, and they do seem to perform a lot better. Most of their expenditures are above $11,000 per student. Randolph field, which is the best performing district, even has expenditure per pupil above $14,000. However, these three are all military base school districts, which might be the reason for their possibility of high expendi- ture. For other school districts, they might not be able to get such high funding, so these three might just be spe- cial cases

Therefore, we conclude that our model shows that increasing funding does not necessarily have a positive im- pact on student performance, and there is not a very consistent relationship between student performance and in- creasing funding. However, three school districts show that when the expenditure per pupil reaches a certain high level, it does appear to have a very positive impact on student performance.

# 5 CONCLUSIONS

This work contributes to both social network analysis and multi-agent system. We propose to model social net- works using multi-agent systems where agents frequently interact with each other. By implementing real school networks, even with the limited data, this work has shown some promising results. With the hierarchical struc- ture generated, we are able to evaluate funding polices for past three years for Bexar County school districts and conclude that increasing funding does not necessarily have a positive impact on student performance. However for some special types of school districts (such as military based Ft Sam Houston, Lackland and Randolph field), when the expenditure per pupil reaches a certain high level, it does appear to have a positive impact on student performance.

We use the real data of 15 school districts in Bexar County, Texas. The publicly available data has some li- mitations that consequently narrow our evaluations. First our current experiment only contains major actors re- levant to funding, such as students, teachers and administrators. It does not include many other effective actors such as parents, vendors, etc. Also our evaluations are only among public schools because of the availability of their data. It will be more interested if the comparisons can be extended to public schools, private schools and charter schools, and so the decision model is not only be used to evaluate funding policies but also school choices among the three types of schools.

In the funding policy evaluation part, the current reward function is unified for all sub-tasks because we only have data for student performance of the whole school district. The ideal situation would be to have performance data on each sub-task, e.g. regular education, career, special education, ESL (English as Second Language), and gifted education. Therefore the funding policy evaluation could be used to analyze the impact to each sub-group rather than just to the whole.

Last but not least, the reward function could be re-designed to cover more comprehensive evaluations to school quality. The current policy evaluation model is just based on student scores. Several researchers and or- ganizations have published papers and reports regarding factors or indicators related to school quality (US De- partment of Education; The Organization for Economic Co-operation and Development). Many of them sug- gested that besides student scores on standardized achievement tests, there are many other performance meas- ures should also be considered, such as teacher retention, teacher transfer rate, school safety, financial invest- ments, school climate, etc. [36] . And in addition to using quantitative measures, [36] also recommends using educational experts to make qualitative assessments of a school.

## ACKNOWLEDGEMENTS

Jiang Wu is supported by The Fundamental Funds for Central Universities, China.

## REFERENCES

1. Andrighetto, G., Boella, G., Sichman, G. and Verhagen, H. (2009) Social Networks and Multi-Agent Systems Sympo- sium (SNAMAS-09) Introduction. Proceedings of the SNAMAS Symposium, Edinburgh, 6-9 April 2009, 1-3.
2. Newman, M.E. (2003) The Structure and Function of Complex Networks. SIAM Review, 45, 167-256.
3. Guessoum, Z. (2003) Dynamic and Adaptive Replication for Large-Scale Reliable Multi-Agent Systems. In: Garcia, A., Lucena, C., Zambonelli, F., Omicini, A. and Castro, J., Eds., Software Engineering for Large-Scale Multi-Agent Sys- tems, Springer, Berlin, 182-198.
4. Grant, T. (2009) Modeling Network-Enabled C2 Using Multiple Agents and Social Networks. Social Networks and Multi-Agent Systems Symposium (SNAMAS), 6-9 April 2009, Edinburgh, 30-35.
5. Ma, J., Guo, D.W., Wang, K.P., Liu, M. and Chen, S. (2009) Colony Evolution in Social Networks Based on Multi- Agent Systems. IEEE International Conference on Natural Computation, 4, 594-597.
6. Bettinger, E. (2005) The Effect of Charter Schools on Charter Students and Public Schools. Economics of Education Review, 24, 133-147.
7. Lubienski, C. and Lubienski, S.T. (2006) Charter, Private, Public Schools and Academic Achievement: New Evidence from NAEP Mathematics Data. Occasional Paper, A12.
8. Slate, J.R. and Jones, C.H. (2005) Effects of School Size: A Review of the Literature with Recommendations. Essays in Education, 13, 1-22.

9. Schaeffer, S.E. (2007) Graph Clustering. Computer Science Review, 1, 27-64. Harris, D.N. and Sass, T.R. (2011) Teacher Training, Teacher Quality and Student Achievement. Journal of Public Economics, 95, 798-812.

10. Clark, D., Martorell, P. and Rockoff, J. (2010) School Principals and School Performance. National Center for Analy- sis of Longitudinal Data in Education Research, Technical Report, Columbia University, New York.

11. Meier, K.J., O'Toole Jr., L.J. and Goerdel, H.T. (2003) School Superintendents and School Performance: Quality Mat- ters. Texas Educational Excellence Project, Technical Report.

12. Anderson, N. (2011) Per Pupil Spending: How Much Difference Does a Dollar Make? All Graduate Plan B and other Reports, Paper 20.

13. Crampton, F.E. (2009) Spending on School Infrastructure: Does Money Matter? Journal of Educational Administration, 47, 305-322.

14. Fortunato, S. (2010) Community Detection in Graphs. Physics Reports, 486, 75-174.

15. Wakita, K. and Tsurumi, T. (2007) Finding Community Structure in Mega-Scale Social Networks. Proceedings of the 16th International Conference on World Wide Web, Banff, 8-12 May 2007, 1275-1276.

16. Dietterich, T.G. (2000) An Overview of MAXQ Hierarchical Reinforcement Learning. In: Koenig, S. and Holte, R., Eds., Abstraction, Reformulation, and Approximation Anonymous, Springer, Berlin, 26-44.

17. Kenyon, D.A. (2007) The Property Tax-School Funding Dilemma. Lincoln Institute of Land Policy, Cambridge.

18. Waters, J.T. and Marzano, R.J. (2006) School District Leadership That Works: The Effect of Superintendent Leadership on Student Achievement. Mid-Continent Research for Education and Learning.

19. Dhuey, E. and Smith, J. (2011) How Important Are School Principals in the Production of Student Achievement? Wor- king Paper, University of Toronto, Toronto.

20. Rockoff, J.E. (2004) The Impact of Individual Teachers on Student Achievement: Evidence from Panel Data. American Economic Review, 94, 247-252.

21. Wong, K.K. (1989) Fiscal Support for Education in American States: The "Parity-to-Dominance" View Examined. American Journal of Education, 97, 329-357.

22. Zhang, D. (2008) The Effects of Teacher Education Level, Teaching Experience, and Teaching Behaviors on Student Science Achievement. Graduate Thesis, Utah State University, Logan.

23. Hanushek, E. and Rivkin, S. (2007) Pay, Working Conditions, and Teacher Quality. The Future of Children, 17, 69-86.

24. Croenweth, S. (2012) Does Teacher Performance Pay Improve Student Achievement?

25. Bohte, J. (2002) School Bureaucracy and Student Performance at the Local Level. Public Administration Review, 61, 92-99.

26. Epple, D. and Romano, R.E. (1998) Competition between Private and Public Schools, Vouchers, and Peer-Group Ef- fects. The American Economic Review, 88, 33-62.

27. Henderson, A. (2002) A New Wave of Evidence: The Impact of School, Family, and Community Connections on Stu- dent Achievement. Southwest Educational Development Laboratory, Annual Synthesis.

28. Ferreyra, M. and Liang, P. (2011) Information Asymmetry and Equilibrium Monitoring in Education. Journal of Pub- lic Economics, 96, 237-254.

29. Chaney, B., Burgdorf, K. and Atash, N. (1997) Influencing Achievement Through high School Graduation Require- ments. Educational Evaluation and Policy Analysis, 19, 229-244.

30. Marchiori, M. and Latora, V. (2000) Harmony in the Small-World. Physica A: Statistical Mechanics and Its Applica- tions, 285, 539-546.

31. Znaniecki, F. (1986) Social Relations and Social Roles. Irvington Publishing, Manchester.

32. Müllner, D. (2011) Modern Hierarchical, Agglomerative Clustering Algorithms. ArXiv Organization, Technical Report.

33. Barto, A. and Mahadevan, S. (2003) Recent Advances in Hierarchical Reinforcement Learning. Discrete Event Dyna- mic Systems, 13, 341-379.

34. Mirzazadeh, F., Behsaz, B. and Beigy, H. (2007) A New Learning Algorithm for the maxq Hierarchical Reinforcement Learning Method. International Conference on Information and Communication Technology (ICICT'07), Dhaka, 7-9 March 2007, 105-108.

35. Pinkus, L. (2009) Moving Beyond AYP: High School Performance Indicators. Alliance for Excellent Education. Policy Brief, Washington DC.

# 6

# CHAPTER

# MULTI-AGENT ARCHITECTURE FOR THE DESIGN OF WSN APPLICATIONS

**Abdelhakim Hamzi[1], Mouloud Koudil[1], Jean-Paul Jamont[2], Michel Occello[2]**

[1]*Ecole Nationale Supérieure d'Informatique, LMCS, Alger, Algeria*

[2]*University of Grenoble, LCIS/INPG-UPMF, Valence, France*

## ABSTRACT

Complex and distributed systems are more and more associated with the application of WSN (Wireless Sensor Network) technology. The design of such applications presents important challenges and requires the assistance of several methodologies and tools. Multi-Agent systems (MAS) have been identified as one of the most suitable technologies to contribute to this domain due to their appropriateness for modeling distributed and autonomous complex systems. This work aims to contribute in the help of the design of WSN applications. The proposed architecture exploits the advantages of MAS for modeling WSN services, network topologies and sensor device architectures.

**Keywords:** Wireless Sensor Networks; Distributed System Design; Multi-Agents System; Agent Metamodel

# 1. INTRODUCTION

Wireless sensor networks (WSNs) are emerging as powerful platforms for distributed embedded computing supporting a variety of high-impact applications such as disaster/crime prevention and military applications, environmental applications, health applications, and smart spaces [1]. These networks pose important challenges for engineers working in the development of such systems: potential high number of nodes, limited resources, sensor heterogeneity, unreliable networks of changing topology, several levels of data processing [2].

In the last decade, wireless sensor networks have been interested by lot of researchers, the result is a huge number of works covering sensor devices design, routing and MAC protocols implementation, energy management, operating system and middleware, simulation, security, localization, deployment, etc. Despite these efforts, this domain still presenting important challenges and needs more potential focus, especially in methodologies and architectures helping the design of complex distributed systems and applications for WSNs [3].

Multi-Agent systems (MAS) have been identified as one of the most suitable technologies to contribute to this domain due to their appropriateness for modeling distributed and autonomous complex systems. In this paper, we propose a multi-agent architecture for the design of WSN applications where we tried to cover almost services of such system taking in consideration the heterogeneous configuration of the network. First, we will present the characteristics, requirements and design process of WSN applications with some related works using MAS paradigm. Before detailing our architecture and the agents composing its different modules, a WSN nodes types and agents classification is necessary to understand the proposed models. Finally, we will finish by a conclusion and some perspectives.

# 2. BACKGROUND

## 2.1. WSNs Characteristics

WSNs are usually wireless dynamic networks composed of a large number of diverse nodes generally equipped with autonomous devices with continuous sensing, data processing and communication capabilities. These nodes are densely distributed in some area without human participation. Each sensor node is capable of detecting environmental conditions such as sound, temperature, and the presence of certain objects.

Although WSNs share common problems and solutions with embedded systems and ad hoc networks in what concerns hardware and network issues, there are new challenges that rise from the fact that these devices are resource constrained in term of energy, processing, storage and communication capabilities.

**Table 1** presents a comparison between Wireless Sensor Network technology and other monitoring technologies [4].

The research community has identified sensor networks as a very challenging domain because of the following distinguishing features [4]:

**Table 1**. Differences between WSNs and other technologies.

| Wireless Sensor Networks | Alternative Technologies |
|---|---|
| Low-cost low-power simple sensors | Expensive high-power consuming complex sensors |
| Cover wide-range areas | Cover small-size areas |
| Monitor remote or hostile environments | Monitor highly-controlled environments |
| Fault-tolerance and robust to node failures | Non-robust |
| Non-invasive | Invasive |
| Irregular sampled datasets | Regularly sampled datasets |
| Intrinsic distributed structure | Intrinsic centralistic structure |
| Low-bandwidth connectivity | High-bandwidth connectivity |
| Battery-powered | Electric-powered |

Complexity: there is no easy way to manually design a sensor network that acts properly in all possible environmental and network changes.

Scale: they are usually composed of thousands of nodes, making infeasible approaches where the computational cost is exponential to the number of sensors.

Physical distribution: during their operation sensors have to deal with computation and information sources that are physically distributed.

Dynamics: sensor networks are dynamic systems that by effect of its internal changes or by effects of external forces change over time.

Resource availability: a key characteristic of sensor networks is that the demand for resources such as computation, power or bandwidth, is always higher than supply. Thus, there is a need for resource-awareness at all operation levels of the network.

Interdependence: the network may need to coordinate different sensors to achieve high-level tasks. Therefore, there are dependencies among sensors that make necessary to coordinate them during the sensor network operation.

Situatedness: Sensor networks are usually located in rapidly changing environments where the decision-making process of the sensor network has severe time restrictions.

## 2.2. WSNs Requirements

Any WSN system or application needs to response to some requirements inherent to its functioning such as nodes hardware and software requirement, deployment, communication, and energy management. Additional requirement, like QoS (Quality of Services) and security are sometimes necessary to guarantee an efficient operation of the WSN. Each requirement has it own parameters which can differ from an application to another.

### 2.2.1. Devices Parameters

The software and hardware parameters of sensor nodes are very important for the development of any WSN system. The hardware parameters include hardware platform specification (memory, chip, interfaces, radio, GPS, actuator, and energy resources) and sensor platform descriptions. The software parameters include the required OS/Virtual machine, needed amount of memory (RAM/ ROM), a list of modules that the considered software module requires to operate and requirements to the hardware, like a radio or a specific sensor support.

### 2.2.2. Deployment or Network Parameters

Network parameters allow to describe the sensor network where the application is deployed. A good sensor network deployment should address a variety of problems such as sensing coverage (Topology, devices heterogeneity, density, average network diameter) network connectivity (bandwidth, mobility, fault rate), and deployment method (deterministic or ad hoc, terrain specifics, Moving object characteristics). Another deployment objective is the trade-off between the network lifetime and the number of sensor nodes (coverage, connectivity and redundancy).

### 2.2.3. Communication

Three major requirements are the target of every network protocol design: bounded delay, energy awareness and low overhead imposed to the network. These three requirements however seem to be contradictory to each other, thus demanding a trade-off in order to enhance one aspect of network performance at the expense of the rest.

Network Layer: Routing protocols are basically grouped by the proactive or reactive way they create routes. No standard metric exists, since each application imposes different requirements on the routing protocol. Some approaches relate this metric to energy consumption and some others to the real-time performance (hop-count, bandwidth).

MAC Layer: MAC protocols cover the handling of collision, minimizing the overhead and synchronization between nodes (idle, listening).

### 2.2.4. Energy Management

Due to the energetic resources restriction, both local and global power consumption management is necessary in order to increase the lifetime of the network. Each node can locally control its devices operations (sensing, processing, communication, etc.) according to the present energy. A good tasks scheduling must take in consideration the energetic capability of each node.

### 2.2.5. Query Management

Since the WSN is used to response to different kind of systems and applications requests, an efficient scheduling of queries is necessary. Some parameters should be taken in consideration when constructing queries.

such as: sources of data, periodic or instantaneous gathering, deadline, aggregation function, etc.

### 2.2.6. Quality of Services (QoS)

QoS refers to the capability of a network to provide better service to selected network traffic over various technologies. Delay and fault-tolerance are two main QoS parameters of wireless sensor networks. The first specifies the maximum allowable propagation delay between the gateway (base station or sink) and the farthest node from it, and the second is the tolerance of the WSN to node failure. A compromise between QoS requirements and WSN cost is necessary according to the type of the application.

### 2.2.7. Security

All security mechanisms are based upon cryptographic algorithms, discriminated in two major categories: private (symmetric) and public (asymmetric). Private cryptography is less demanding on computational power, but this imposes considerable control overhead. Public cryptography solves the key number issue, but it affects negatively the node performance and results into larger cipher data, burdening memory usage. Efficient management of the keys is crucial for WSN security architecture because of the ad hoc topology of the WSN and the restriction of nodes' resources.

## 2.3. Designing WSNs

Designing and programming WSNs is a complex task as system level code and application level code are often intertwined (see **Figure 1**). The users (and developers) of WSNs are often not software engineers but domain experts in other fields, who use WSNs to obtain data relevant to their work. Although this disparity in expertise, they are required to operate in a concurrent, realtime, resource constrained computing environment that is potentially complex and hard to program correctly. Several programming approach have been used in order to facilitate this task such as Object-Oriented [5], Component-Oriented [6], Service-Oriented [7] and Agent-Oriented programming [8].

The concept of agent provides a convenient and powerful way to describe a complex software entity that is capable of acting with a certain degree of autonomy in order to accomplish some tasks. The agent can be defined as autonomous, problem solving computational entities capable of effective operation in dynamic and open environments like WSNs. Typical characteristics of the agents are autonomy, intelligence, mobility, and social ability. The agents can be classified into multi-agent and mobile agent. The multi-agent is for handing complex operations requiring collaboration between the agents. The mobile agent moves through the network to process the tasks, and it is widely used for mobile computing with wireless network [4].

Various agent-oriented methodologies, metamodels and architectures exist to design and develop distributed and complex systems in an abstract manner. Regarding to WSN, almost contributions focus on the use of mobile agents to solve a particular problem such as routing, clustering, deployment, localization and security [9]. To make programming mobile agent for WSN easier, some architecture was proposed. The most popular are Sensorware [10] and Agilla [11].

Some interesting architectures tried to cover the gap in autonomous systems and WSN design. For example [12] is a tailored approach using MAS for the design of embedded collective systems and have been validated with the development of WSN application. [13] discuss a design methodology based on the service-oriented architecture and agile development principles for wireless embedded and sensor networks. [14] presented an agentbased framework acting as an integral part of a middleware to support autonomous setup and adaptation of sensor networks. [15] tried to exploit the advantages of model-driven engineering (MDE) approach to propose an architecture for the design of WSN based on the INGENIAS modeling language. [16] proposed a specification language centered on the concept of lightweight agent based on events and states. [17] proposed an interoperable model provided generalized frameworks to address some critical issues existed in the WSN such as interoperability, time synchronization, power management and distributed computation.

Frequently, almost proposed architectures and frameworks specialize on particular parts of WSN design and applications and only few works have been invested to derive a general architecture. This field still needs more investigation. Our work fills in this objective. It defers from related works in term that it tries to cover almost WSN's application requirements and different kind of network topology and nodes architecture.

# 3. AGENTS DEFINITION CRITERIA

There are two main parameters that define the role of the agent in the system: the type of the node with the role it plays in the WSN and the type of the agent itself.

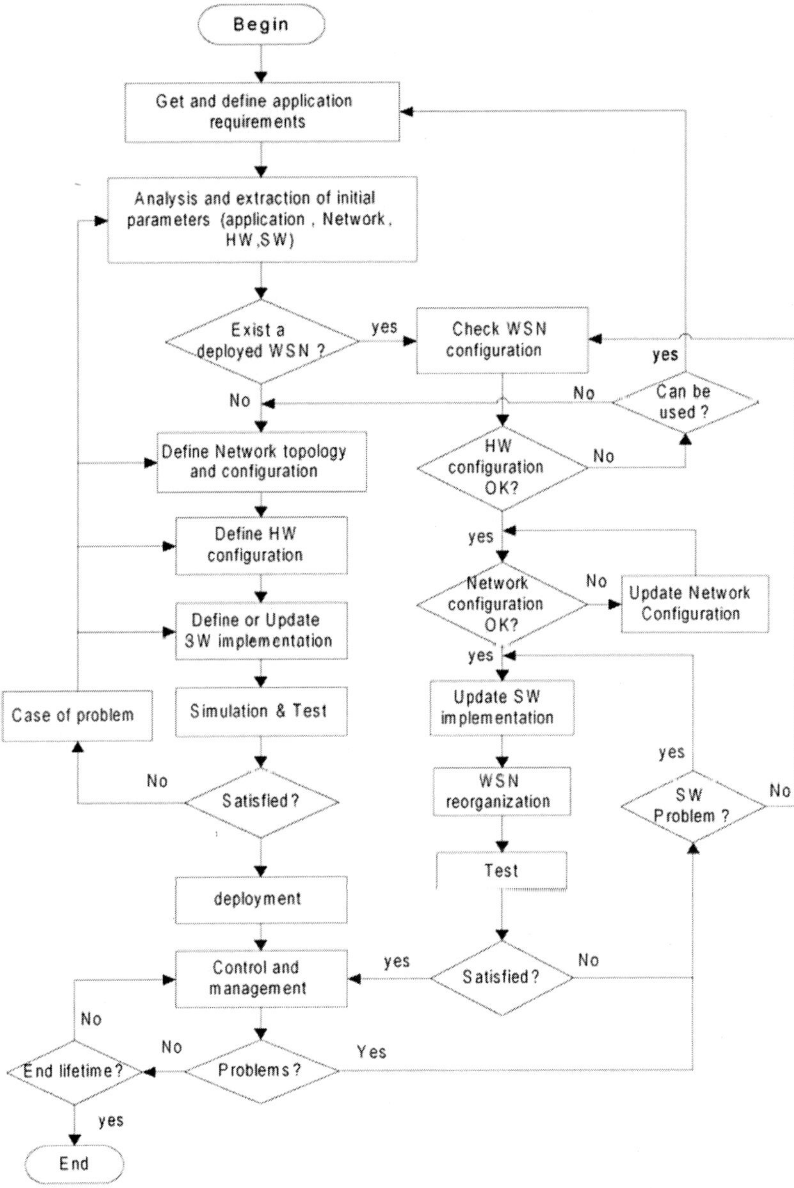


**Figure 1**. WSN applications design process.

## 3.1. WSN Nodes Classification

The type of the node is strongly related to its hardware components which specify its functions in the network. Some functions are done by all nodes, the communication for example. Which define more the type of a node are the additional functions that this node accomplishes.

- Base Station (or Sink) (BS): a well equipped node with more processing and storage capabilities. It is responsible of queries dissemination and data gathering. It can perform some processing on the data before being transferred. We can have more than one Base Station and some of them can be mobile.

- Communication Node (CN): generally, all the nodes can participate in the communication; some nodes can have other roles such as: sensing, processing and storage. The nodes whose role is only the communication are designed as CN.

- Sensor Node (SN): this type of node is responsible of the capture of measures and transforming them to a digital form to be sent to other nodes. Commonly used sensors include temperature, pressure, humidity, vibration, etc. Some sophisticated devices can have video camera.

- Processing Node (PN): this type of node has more processing capability than other nodes in order to perform some process on the gathered data such as aggregation, compression, encryption, etc. Generally this kind of nodes is designed as cluster-head.

- Storage (or Memory) Node (MN): generally, these nodes are used as cache memory for the WSN in order to store the data captured from the sensor nodes and not have been yet requested by the Base Station, or the code that may be dispatched to the nodes if they change their roles in the network.

- Actuator Node (AN): this type of nodes has the capability to react with its environment. For example it can trigger an alarm, send a signal to another system or change its position.

- Localization Node (LN): some applications or protocols need to know the position of some nodes in the observed area. There are lot of localization algorithms, but almost of them are based on the knowledge of the position of predefine nodes called anchors nodes. Generally, the positions of these nodes are calculated using a GPS system.

## 3.2. Agents Classifications

Before giving a description of the agents composing the proposed architecture and their tasks in the system, it is important to present a short classification of agents. This can considerably help to understand the different types of agents and their roles. This classification is based on the metamodel proposed in [11] with some adaptation. In general, almost agent's types can be classified according to two views: mobile/fixed and reactive/cognitive.

### 3.2.1. Mobility

According to this classification, the agents can be fixed in a hosting nodes or have the ability to move from one node to another.

#### 3.2.1.1. Fixed Agent

This kind of agent is an autonomous program installed in the node for high-level directions of missions. It reasons according to the conditions of the system, the requirements of the missions attributed to it, and its own state and that of its hosting node and neighborhood, in order to take a local decision related to its engagement in a given mission. We define two types of this agent:

Task-agent: this agent has limited deliberation capabilities. It performs specific and predefined tasks without make any intelligent decision. Generally, these tasks are controlled by a planning-agent.

Planning-agent: this agent has intelligent behavior. It is used to complete more complicated mission. Its reasoning and decisions are based on the changes of the state of the component where it is installed as well as on the predefined global-mission of this component.

#### 3.2.1.2. Mobile Agent

A mobile agent is an autonomous program that moves from a node to another one, so it can run itself as soon as it needs without being called by other routines. It can determine where to migrate autonomously and carries on running its own program according to the previous state transferred from the last node where it existed. Mobile agents are classified according to their usage, as presented below:

Service-agent: this agent is used to provide a node specific processing services that is not yet installed, such as encryption, compression, and mathematical functions. Such agent have limited reasoning capabilities. Its intelligence is limited to the decision to move to another node or to clone itself and send its clones to some other nodes in response to an incoming message or internal event.

Update-agent: this agent provides changes in the components/agents installed in the system. It is used to perform administrative tasks, such as software updates. This kind of agent is temporary and it is excluded from the node just after finishing its updating task.

Mission-agent: this agent is responsible for carrying, disseminating and allocating node-missions to different nodes in the network. Its role is to take the result of the global-mission interpretation and to carry it to the nodes in the network, according to the location specified in the mission directions.

### 3.2.2. Computational Model

#### 3.2.2.1. Reactive

Agents Service-agents, update-agents and task-agents have limited deliberation capabilities. Their behavior is basically modeled by a direct reply to a certain messages or events, without any need to perform elaborated reasoning about any statements or parameters interpretation. This fact leads to a reactive-based computational model, where the specific reactions depend on their internal mental state. Generally these agents run by the control of the planning-agent to perform some actions such as: execute-task, move, clone, install, send, and receive (request and reply) messages to and from other agents.

#### 3.2.2.2. Cognitive Agents

Mission-agents and planning-agents have more sophisticated intelligent behaviors. In order to perform their activities, they require some cognitive skills. The cognitive agent model adopted is based on the BDI (Beliefs-Desires-Intentions).

Mission-agent: This agent has to take different decisions during its lifetime, such as: the number of clones it will make from itself in order to disseminate a given mission; the type of node-mission to deliver to each node; and whether to deallocate itself from a node. These kinds of actions require mission-agents to communicate with the planning-agent that governs the hosting node to get information about what to do under given circumstances.

Planning-agent: The planning-agent has the most complex "mental" activity, being responsible for different types of reasoning related to the mission accomplishment. It communicates with all other kinds of agents. This agent is responsible for deciding if the node will take or not a given mission. It also has to maintain updated information about its own state, in order to inform the other planning-agents and be capable of taking right decisions. Environment conditions are also important in some of the deliberations taken by this agent.

# 4. THE ARCHITECTURE OF THE FRAMEWORK

The proposed architecture is composed of functional components or modules; each one is a composition of agents ensuring the services offered by this component as shown in the **Figure 2**.

## 4.1. Modules Description

In this section we will describe the main modules without entering in the role of the different agents used by these modules. More details are presented in the next section.

- System Module (SM): it presents the kernel of the application. The main role is to group the other modules and guarantee a coherent communication between them firstly, and with the WSN in the other hand.
- User Interface Module (UIM): the data collected by the WSNs can be directly used by the end user or by another system that makes some process on the data before being useful, or react according to them. This module presents an interface of communication between the WSN system and the user of the data gathered. This user can be locally connected to the WSN system or using a communication media. Several interface agents may exist because the WSN can be used by many different users (systems or applications) in the same time.
- Data Processing Module (DPM): generally the data collected from the area where the WSN is deployed need to be processed before being presented or communicated to the user. In some cases, the system using the WSN receives requests system or applications. Sometimes it is necessary to process and interpret these requests before sending them to the adequate WSN (the system can manage several WSNs).
- System Management Module (SMM): this module manages the functions of the system modules and ensures their good operability with the WSNs.

- Network Interface Module (NIM): this module is the interface between the system and the different WSNs. The SMM uses the NIM to send requests to the WSNs and receive data from them. It also allows the SMM to manage and control the different WSNs via the Network Management Modules.
- Network Management Module (NMM): the system can use and manage several WSNs in the same time for different purposes. Each WSN has its own NMM that ensures its functions and communicates with the system via the NIM under the control of the SMM. The main role is the management of the functional integrity and a coherent communication between the WSN modules.

## 4.2. Modeling Agents

According to the modules composing the system, the services needed by the WSN, the structure and the roles of the nodes, several agents may exist. These agents respect the classification presented above in order to specify their roles in the system and the network. Using these agents' models in our architecture, we tried to cover almost WSN applications, network topologies and protocols, and nodes architectures.

### 4.2.1. Interface Agents

The system managing the WSNs can be connected to several users or systems using different media. For each class or type of interface we can have a task-agent that manages this interface. Because these agents can use different means of communication to simply send information produced from the WSNs or perform some reactions, we can have more than one communication-agent of mission-agent and task-agent type.

### 4.2.2. System Management Agent

This agent is the principal controller of the system. It is a planning-agent that reasons according to the state of the system and the WSNs managed by this system as well as the requests received throw the user interface to make the adequate decisions and order the other agents of the system to perform some tasks.

### 4.2.3. Data Processing Agents

The system can perform some additional process on the data gathered by the WSNs (analysis, compression, encryption, etc.) before being sent to the user or for helping the System Management Agent to make the right decisions. For each kind of processing we can have.a taskagent.

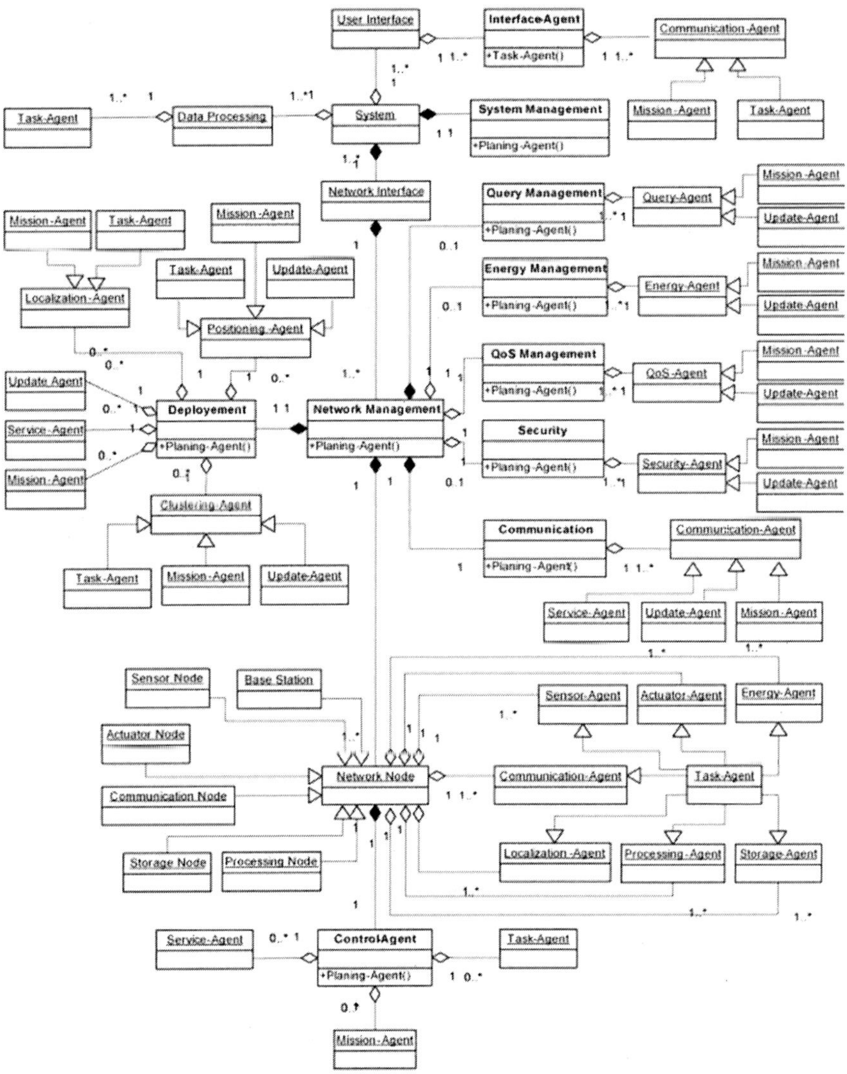


**Figure 2**. Framework architecture.

### 4.2.4. Network Management Agents

The system is interfaced with one or more WSNs using the network interface module. For each WSN there is a planning-agent that controls this network by ensuring a functional integrity and coherent communication between the different agents, and makes its decisions according to the state of the managed WSN and the global mission given by the System Management Agent. The principal roles of these agents are: ensuring an efficient deployment, organization and control of the WSN nodes; preparing queries and receiving responses; managing communications, energy consumption, QoS and security.

### 4.2.5. Deployment Agents

These agents are responsible of the deployment, reor ganization and update of the WSN. To do this, a planning-agent that has knowledge about the state of the network and all nodes, control other agents to perform specific tasks such as localization, positioning and clustering. The planning-agent can use other agents for nonpredefined tasks, for example the update-agent can be used to install some software update in the nodes, the service-agent is used to perform processing tasks that other agents need to accomplish their tasks and the mission-agent is used to assign the nodes a specific local missions.

- Localization Agents: the role of a localization-agent is to calculate the geographic position of the nodes. The method of localization can be centralized; in this case it is carried out by a task-agent, as it can be distributed; in this case a mission-agent is disseminated to the nodes responsible of this task.
- Positioning Agents: the role of a positioning-agent is to change the geographic position of the mobile nodes in the network. This can be done by a centralized or decentralized method. In the first case a task-agent choose the position according to the decision of the planning-agent of the deployment module and an update-agent is used to order the mobile node to take the specified position. In the other case, the planningagent disseminates mission-agents to the mobile nodes to collaborate together in order to negotiate and decide their positions.
- Clustering Agents: the clustering in a hierarchical topology network is the reorganization of the WSN on groups of nodes where a cluster-head is elected to manage the group of nodes and perform some processing tasks on the data gathered by its members. The clustering phase is very important for some hierarchical routing protocols, so it is very useful to assign to the WSN such service. Like the positioning, this task can be

centralized or decentralized. In the first case, a task-agent specify the cluster-heads and the members of cluster using the parameters given by the planning-agent of the deployment module, then an update-agent is disseminated to update the role of each node in the network. In the other case, the planningagent disseminates mission-agents to the nodes in order to negotiate and elect the cluster-heads and the members of clusters.

### 4.2.6. Query Agents

The query agents are responsible of analyzing the received requests, constructing queries and choosing the adequate base stations that can disseminate them to their closed nodes. They can also manage the storage nodes used as cache memory of the captured data. These tasks are managed by a planning-agent which controls other query-agents. A query-agent can be a simple updateagent that updates a specific base station or a storage node with a list of queries, or a mission-agent that is disseminated to the base stations which decide according to their states the capability to perform this mission or to collect some data from storage nodes.

### 4.2.7. Communication Agents

These agents implement the protocols of communication used for the selection of the efficient multi-hop path that data take from the nodes to the base stations. A planning-agent is responsible for the management of the process of communication. It can decide which role to play by each node involved in the communication according to the topology of the network. It also can ask the deployment-agent to reorganize the network when it observes an inefficient routing. The planning-agent controls three other types of agents: the update-agent is used to update the network layer of the nodes with the implementation of the routing protocol used, the mission-agent is used in the case where the routing protocol is based on a mobile agents, and service-agent is used to add to the routing protocol processing services (aggregation, compression, encryption, etc.).

### 4.2.8. Energy Management Agents

Because of the energetic limit of the nodes, the management of energy consumption is a crucial task in a WSN. The energy management agent is a planning-agent that has knowledge of the energetic state of nodes, so it can provide other management agents with some parameters that can help them in their decisions. This planning-agent can control tow kinds of energy-agents: the update-agent updates the energy-agent of the node with the strategy of the consumption of the energy and the mission-agents that can traverse the WSN in order to make knowledge about the energetic state of the network.

### 4.2.9. QoS Management Agents

QoS agents try to guarantee some degree of efficiency for real-time and high-level of availability for the WSN applications. A planning-agent collaborates with other agents (deployment, query, energy, communication) in order to assign certain priories to some nodes. A QoS strategy can be defined by the planning-agent then dispatched to the concerned nodes using an update-agent, or mission-agents are disseminated to a specific area, analyze the states of the nodes and their environment then negotiate the adequate strategy of the QoS.

### 4.2.10. Security Management Agents

The main roles ensured by these agents are: public key management, authentication, data encryption and malicious nodes detection. To do these, two kinds of agents can collaborate under the control of the planning-agent. The update-agent assigns the nodes with security policy and parameters defined by the planning-agent. The mission-agents are used in the case where the security mechanism use mobile agent for example to detect security vulnerability in the network or malicious nodes.

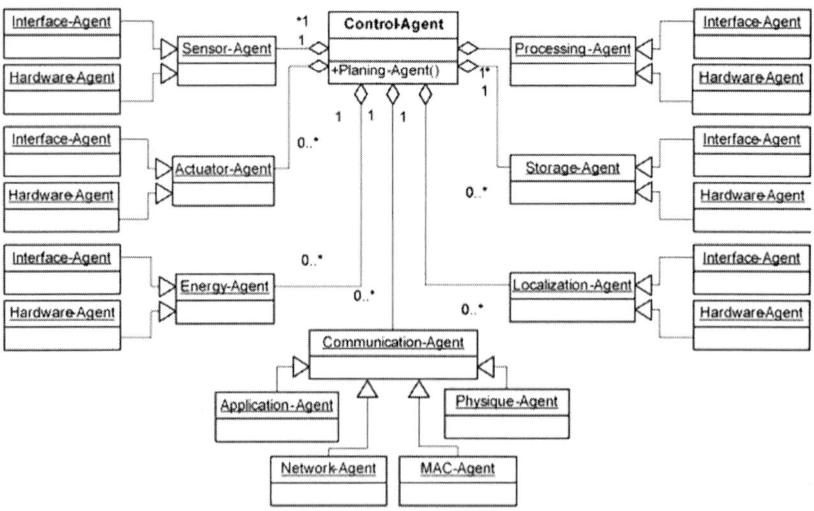


**Figure 3**. Different types of nodes agents (Hardwareand Interface-agents are Task-agent type).

### 4.2.11. Nodes Agents

Depending on the hardware components of the node and the role it plays in the network, several agents can exit as shown in **Figure 3**.

- Control Agent: this agent is the coordinator of the node. The main roles are the control of its functions, the collaboration with the control agents of other nodes, the reception of updates and missions, and the coordination between the other agents present in this node. The type of this agent is planning-agent so it has reasoning and decision capabilities based on the global-mission of the WSN, the local-mission assigned to this node by the Network Management Agent, the message received from other nodes and the event trigged by the hardware component controlled by the other local agents.

- Communication Agents: these agents are responsible of the communication with the other nodes. Because of the communications task is based on layered model, for each layer there is task-agent that ensure the tasks done by this layer under the supervision of the control agent. For example the Physique-agent control the state of the transceiver (idle, transmission, transmission-reception, signal power, etc.), the MAC-agent ensure the timing and synchronization between nodes, the Network-agent implement the routing protocol tasks and the Application-agent perform some processing on the data received from another nodes.

- Sensor Agents: Sensor Agents are task-agent. They manage all kinds of sensors present in the nodes. There are two kinds of agents: the sensor-hardwareagents which control the function and the state of the sensors (idle, acquisition, sensing parameters, etc) and the sensor-interface-agent which play the role of interface between the Control Agent and the sensors installed in the nodes.

- Energy Agents: the energy in wireless sensor networks is a crucial constraint. The energy source is classified into three categories: storage (batteries or capacitors), Energy harvesting mechanisms (photovoltaic materials, ambient vibration, thermal gradients), and transfer (inductive coupling, capacitive coupling, and passive backscattering) [18]. The role of energy agents is the control of the energy generation and consumption and informing the control agent with the state of the energetic capacity of the node. There are two kinds of agents: the energy-hardware-agents and the energy-interface-agent.

- Processing agents: these agents control the processing devices of the node (processor, FPGA, DSP, etc.). Like other hardware components, there are two kinds of agents: processing-hardware-agent and processinginterface-agent.

- Storage Agent: this agent is responsible for the management of data and code storage. It collaborates with other storage agents in order to ensure an efficient redundancy and distribution of data in the storage nodes present in the network.
- Actuator Agents: the role of this type of agents is the control of the devices that allow the nodes to react with the environment under the supervision of the control agent (trigger an alarm, send a signal to other devices, and change the position of the node). Like other hardware components, there are two kinds of agents: actuator-hardware-agent and actuator-interface-agent.
- Localization Agents: the role of this type of agent is the control of the localization system like a GPS. There are two kinds of agents: Localization-hardware-agent and Localization-interface-agent.

# 5. EXAMPLE: BATTLEFIELD SURVEILLANCE

## 5.1. Description of the Application

This application example deals with the design of a system which can detect and classify multiple targets (vehicles and troop movements).

The objectives are: classifying objects, estimating the number of each type of objects, estimating the direction and the speed of movements.

1) The objectives require that Software must be able to:

- Classify targets from the power of vibration signal, the estimated distance from some nodes, some view of image and a database of predefined object characteristics.
- Estimate the number of objects, their direction and the speed of movement from the power of vibration signal, the position and the time when the target has been in this position, and the type of targets.

2) We can make Hardware suppositions:

- We have heterogeneous nodes.
- The majority of nodes are sensor nodes using vibration to detect the target.
- Some mobile nodes which can fly for a small distance (hundred of meters). These flying nodes are enough small that can't be detected. They are equipped additionally to the vibration sensor with low resolution camera and GPS, and can update its power from an energy storage station.
- Some data storage nodes used to store the history data of detected target.
- Some small energy storage stations are used to update the energy of the flying nodes.

## 5.2. Modeling the Application

### 5.2.1. Data Structures

- **Query format:**

1) Vibration sensor nodes: [Node_Id, position (x, y), signal_value, direction, time].

2) Camera sensor nodes: [Node_Id, position (x, y), image_data, time]

- **Agent format:**

(Node_Id, Module_Id, Agent_Id, Role_Id, [List_Task], [List_Object]).

Node_Id: the identifier of the hosting node.

Module_Id: the identifier of the module or the component where the agent is executed. If we have submodules, the identifier of the internal module is the concatenation of all its encapsulating modules identifiers.

Agent_Id: the identifier of the agent. According to the identifier we can know the type of the agent. For example the identifiers of planning-agents is between 0 and 31 of the mission-agent between 32 and 127, and so.

Role_Id: the identifier of the role assigned to the agent. To define the code executed by the agent we use three tables:

1) Roles table: each entry has two information: the Role_Id and list of Task_Id.

2) Tasks table: each entry has two information: the Task_Id and list of Object_Id.

3) Objects Table: each entry has three information: Object_Id, parameters and Object_Code. The parameters is inherent to object execution such as the memory size necessary for the execution of the Object_Code, the energy consumption per time unit by the processor, and energy consumption per time unit by the transceiver (Send_Energy and Receive_Energy).

### 5.2.2. Definition of Agents

We will define some important agents as illustration.

- **Query and processing agents**

They request and analyze data in order to classify targets and estimate their number and movements. **Table 2** gives examples of query and processing agents and their roles.

The centralized processing is used in the case where the troop is moving rapidly. It needs a big volume of real-time data so it decreases the lifetime of the network.

For the distributed processing a mission-agent is sent from the base station, traverses storage nodes, analyze data, and identify the requested information. It is used when the troop is in rest in order to minimize the quantity of data transferred to the system.

- **Deployment agents:**

The deployment and the reorganization of the WSN is based in several agents as shown in the**Table 3**.

1) For the vibration sensor node, an ad hoc deployment is used. It is important to identify the optimal number of nodes according to the observed area.

2) For energy and data storage node, a deterministic deployment is used in order to ensure an equitable distribution.

**Table 2**. Query and processing agents.

| Agent | Type | Role |
|---|---|---|
| Query | Planning-agent | Manage queries. |
| Update_Query | Update-agent | Update base stations and storage nodes with the format of requested data. |
| Local_Processing | Task-agent | Centralized processing. |
| Mobile_Processing | Mission-agent | Sent from the base station. Traverse storage nodes analyzing data, and identify the requested information. Used when the troop is in rest in order to minimize the quantity of data transferred to the system. |

**Table 3**. Deployment agents.

| Agent | Type | Role |
|---|---|---|
| Deployment | Planning-agent | Manage the deployment, the localization and the reorganization of nodes. |
| Positioning | Planning-agent | Manage the positioning process. |
| Position_Update | Update-agent | Order a flying node to take a specific position. |
| Positon_Define | Mission-agent | The flying node negotiates and decides the position it takes according to the given mission. |
| Localization | Planning-agent | Manage the localization process. |
| Update_ Localization | Update_agent | Inform nodes with the position of other nodes. |
| Negotiate_ Localization | Mission_agent | Nodes calculate their positions with the help of the flying node which knows its position using a GPS. |
| Clustering | Planning-agent | Manage the clustering process. |
| Update_Cluster | Update_agent | Update the roles of cluster members. |
| Negotiate_Cluster | Mission_agent | Decentralized clustering. |

3) For flying node, a deterministic deployment according to the position and the capacity of the energy storage node. Using this kind of node ensure a fault tolerance because the flying nodes can occupy the gap caused by nodes failure meanwhile the reorganization of the network.

4) A hierarchical topology is suitable for the detection of the objects their direction and speed.

# 6. CONCLUSION

Despite the many and significant contributions to the different research topics in WSNs, more research effort is required especially for design aided architectures. The presented approach is intended to facilitate the high-level design of sensor networks based on MAS. It includes an agent-oriented modeling of WSN services and sensor devices architectures. The high degree of reusability of models and transformations, and working at a higherlevel of abstraction than that of code, are expected to reduce the costs of developing the complex autonomous embedded systems. As future works, we will try validate our architecture with the implementation of a real case study application using the proposed models then implementing a middleware based on this work for the automatic (or semi-automatic) generation of low-level code according to the simulators as well as the physical architecture used for the development of WSNs.

# REFERENCES

1. E. Yoneki and J. Bacon, "A Survey of Wireless Sensor Network Technologies: Research Trends and Middleware's Role," Technical Report, UCAM-CL-TR-646, University of Cambridge, Cambridge, 2005.

2. I. F. Akyildiz, W. L. Su, Y. Sankarasubramaniam and E. Cayirci, "A Survey on Sensor Networks," IEEE Communications Magazine, Vol. 40, No. 8, 2002, pp. 102-114.

3. J. Yick, B. Mukherjee and D. Ghosal, "Wireless Sensor Network Survey," The International Journal of Computer and Telecommunications Networking, Vol. 52, No. 12. 2008, pp. 2292-2330.

4. M. Vinyals, J. A. Rodriguez-Aguilar and J. Cerquides, "A Survey on Sensor Networks from a Multi-Agent Perspective," Proceedings of the 2nd International Workshop on Agent Technology for Sensor Networks, Estoril, 13 May 2008, pp. 1-8.

5. Ž. Živanov, P. Rakić and M. Hajduković, "Wireless Sensor Network Application Programming and Simulation System," Computer Science and Information Systems, Vol. 5, No. 1, 2008, pp. 109-126.

6. A. Dearle, D. Balasubramaniam, J. Lewis and R. Morrison, "A Component-Based Model and Language for Wireless Sensor Network Applications," 32nd Annual IEEE International on Computer Software and Applications, Turku, 28 July-1 August 2008, pp. 1303-1308.

7. E. Meshkova, J. Riihijarvi, F. Oldewurtel, C. Jardak and P. Mahonen, "Service-Oriented Design Methodology for Wireless Sensor Networks: A View through Case Studies," IEEE International Conference on Sensor Networks, Ubiquitous, and Trustworthy Computing, Taichung, 11- 13 June 2008, pp. 146-153.

8. Z. Guessoum and J.-P. Briot, "From Active Objects to Autonomous Agents," IEEE Concurrency, Vol. 7, No. 3, 1999, pp. 68-76.

9. M. Chen, S. Gonzalez and V. C. M. Leung, "Applications and Design Issues for Mobile Agents in Wireless Sensor Networks," Wireless Communications, Vol. 14, No. 6, 2007, pp. 20-26.

10. A. Boulis, C.-C. Han, R. Shea and M. B. Srivastava, "Sensor Ware: Programming Sensor Networks beyond Code Update and Querying," Pervasive and Mobile Computing, Vol. 3, No. 4, 2007, pp. 386-412.

11. C.-L. Fok, G.-C. Roman and C. Y. Lu, "Agilla: A Mobile Agent Middleware for Self-Adaptive Wireless Sensor Networks," ACM Transactions on Autonomous and Adaptive Systems, Vol. 4, No. 3, 2009, Article No. 16.

12. J. P. Jamont and M. Occello, "Designing Embedded Collective Systems: The DIAMOND Multiagent Method," Proceedings of the 19th IEEE International Conference on Tools with Artificial Intelligence, Patras, 29-31 October 2007, Vol. 2, pp. 91-94.

13. E. Meshkova, J. Riihijarvi, F. Oldewurtel, C. Jardak and P. Mahonen, "Service-Oriented Design Methodology for Wireless Sensor Networks: A View through Case Studies," IEEE International

Conference on Sensor Networks, Ubiquitous, and Trustworthy Computing, Taichung, 11- 13 June 2008, pp. 146-153.

14. E. P. de Freitas, T. Heimfarth, A. M. Ferreira, F. R. Wagner, C. E. Pereira and T. Larsson, "An Agent Framework to Support Sensor Networks' Setup and Adaptation," Computer Science and Information Technology, Mragowo, 12-14 October 2009, pp. 619-626.

15. R. Fuentes-Fernández, M. Guijarro and G. Pajares, "A Multi-Agent System Architecture for Sensor Networks," Sensors, Vol. 9, No. 12, 2009, pp. 10244-10269.

16. G. Fortino, A. Garro, S. Mascillaro and W. Russo, "Specifying WSN Applications through Agents Based on Events and States," International Conference on Sensor Technologies and Applications, Valencia, 14-20 October 2007, pp. 463-468.

17. F. Y. Xiong and L. Bai, "Interoperable Wireless Sensor Network Model Using Multi-Agent-Based Middleware," International Symposium on Intelligent Signal Processing and Communication Systems, Chengdu, 6-8 December 2010, pp. 1-4.

18. S. Cheekiralla and D. W. Engels, "A Functional Taxonomy of Wireless Sensor Network Devices," 2nd International Conference on Broadband Networks, Boston, 3-7 October 2005, pp. 949-956.

# 7

# CHAPTER

# REVIEWING MICROGRIDS FROM A MULTI-AGENT SYSTEMS PERSPECTIVE

**Jorge J. Gomez-Sanz [1,*], Sandra Garcia-Rodriguez [1], Nuria Cuartero-Soler [1] and Luis Hernandez-Callejo [2]**

[1] *Faculty of Computer Science, Complutense University of Madrid, Madrid 28040, Spain*

[2] *CIEMAT: Centro de Investigaciones Energéticas, Medioambientales y Tecnológicas, Avenida Complutense 40, Madrid 28040, Spain*

## ABSTRACT

The construction of Smart Grids leads to the main question of what kind of intelligence such grids require and how to build it. Some authors choose an agent based solution to realize this intelligence. However, there may be some misunderstandings in the way this technology is being applied. This paper exposes some considerations of this subject, focusing on the Microgrid level, and shows a practical example through INGENIAS methodology, which is a methodology for the development of Agent Oriented systems that applies Model Driven Development techniques to produce fully functional Multi-Agent Systems.

**Keywords:** microgrid; smart grid; multi-agent systems (MAS); INGENIAS; intelligence

# 1. INTRODUCTION

Following Farhangi [1], when comparing a Smart Grid with the conventional Power Grid at the distribution level, the following extra elements can be highlighted: a sensor and metering network, network nodes with computation capabilities, switches or actuators that allow the grid setup to be changed and the capability of plug in or plug out new devices.

However, adding such elements to any Power Grid does not necessarily make the grid smart. New control algorithms are required that coexist or integrate with standard power management mechanisms. A first idea could be defining management rules for each computing node and indicate, in each role, which actions (switching on an element, for instance) should be executed depending on the current state of the network, which is determined through the sensor network. This idea is not so simple:

- Plug in or plug out new elements in a grid, and their proper management, needs the control system to be aware of what it is connected and which services it could provide. New energy storage systems or new generators can also be considered. However, there can be a wide variety of devices to consider such as new sensors or new computing nodes
- Sensors and other metering systems are not always reliable. A sensor can fail, it could be incorrectly installed or could provide different observations. The intelligence system should deal with these issues. Scaling the problem to hundreds of sensors just makes it harder, but processing data in time to take adequate decisions may not be feasible. Even if decisions related to safety and protection are taken directly in the involved devices by specialized hardware, the need of quickly processing (in time) sensor data to take adequate decisions is still a priority.
- Some elements or entire sections of the grid could stop working. The control system should be able to manage these kinds of situations by proceeding in a proper way to preserve the stability of the grid.

- The Power Grid can face warning situations and therefore automatic protection mechanisms should be considered as well.
- Thus, the literature on this subject transmits the concerns of the authors about sensor integration, proficiency in the data management, awareness, computing distribution, coordination among devices and so on. Any solution to be implemented in a grid that tries to make it smart should provide answers to how such capabilities could be incorporated in a coherent way.

Traditionally, when people talked about "electrical grid" they used to refer to the interconnected transmission system. However, the term "Smart Grid" has been more oriented to the entire electrical system including generation, transmission and distribution. Regarding the distribution system, several efforts target the increase of manageability and efficiency by dividing the smart distribution grid into sub-systems. Such sub-systems are called "Smart Microgrids", or just "Microgrids", and consist of energy consumers and producers at a small scale and are able to manage themselves [2]. Examples for Microgrids may be, for instance, villages, industry sites, university campus, *etc.* Furthermore, a Microgrid can either be connected to the backbone grid or to other Microgrids, or it can run disconnected to the electricity grid in island mode.

Since the distribution system is considered as the largest and most complex part of the entire electrical system [3], an extensive literature is focused on Smart Grids located at this level. For this reason, we want to contribute to this research by identifying the most recurrent problems within these kinds of grids. Furthermore, we suggest an improvement of the multi-agent control system, designed by Pipattanasomporn *et al.* [4], for managing Microgrids.

A first hypothesis of this paper is that an important part of Smart Grid researchers thinks that agent research is a field capable of providing preliminary solutions to the above mentioned issues. Smart Grid-related papers are frequently found, which do assume, like Farhangi [1] and Amin and Wollenberg [5] surveys, that intelligent agents are the right technology for Smart Grids. In fact, the IEEE Power Engineering Society Intelligent Systems Subcommittee created the group the IEEE Power and Energy Society Multi-Agent Systems Working Group (MASWG) to identify and disseminate the details of key technical challenges for the proper use of the technology within the power engineering domain.

Since the initial assumption seems to be valid, the paper focuses on a second hypothesis: the use of agent technology made so far in Microgrids is not following the major agent oriented development trends. The fact that agent technology is relevant does not imply that the problem is solved. Actually, the agent research community has been holding debates to determine which way is the best one to apply this concept in current grids. These concerns have been dealt in part with the advent of Agent Oriented Software Engineering (AOSE). Nowadays, a developer can find many methodologies that could be applied to real problems and that are supported by software tools. The use of such methodologies to model the agent oriented part of a Microgrid could improve their management and the better realization of the Microgrid concept.

This paper contributes with a discussion on how agents can be applied to build Microgrids in Section 2. It enumerates some principles that can aid a developer to apply the agent paradigm in a more effective way. Then, it illustrates the idea with an enhancement made on an existing work from Pipattanasomporn *et al.* [4] in Section 3. This work was chosen because for being a highly cited example of the application of agent technology to Microgrids. The experiment was made using INGENIAS (INGENIería de Agentes Software or its equivalent Software Agents Engineering) Agent Oriented Methodology [6] as mean to achieve a different Microgrid than the original work. A first attempt is shown in Section 3 where it does not incorporate any Microgrid simulation. An expanded version, where a Microgrid is simulated, is presented in Section 4. Afterwards, there are some conclusions extracted from this research.

## 2. INTELLIGENCE IN A MICROGRID

To begin with, addressing the intelligence in a Microgrid implies deciding first when one Microgrid will be considered intelligent. The work from Brooks [7] can answer this: "the intelligence is in the eye of the beholder". Brooks studies how research in Computer Science has influenced in the understanding of what Intelligence means for biological systems. Through observation of dynamic environments and how intelligence could occur in those places, the author concludes that artificial intelligent entities are intelligent because they behave, with respect to an external human observer, accordingly. However, the way this behavior is obtained does not necessarily require complex reasoning engines. As a proof of concept, Brooks [8] built his Subsumption architecture. This architecture was a set of layers made of individual automaton.

Therefore, assuming there is a Microgrid that could behave intelligently according to some external observer, the next step is identifying what behaviors current literature considers as intelligent ones. The way to achieve such behaviors is the subject of the rest of the section.

Farhangi [1], and Amin and Wollenberg [5] state that the intelligence is reached with the incorporation of agents to the grids. This is not an isolated fact. A quick inspection of keywords in Google Scholar returns 19,600 documents having the term Smart Grid, whereas Smart Grid and agent together returns 13,500 hits, which is rather significant. Therefore, there is a concern that agents and Smart Grids are highly related, though how these agents do provide this intelligent behavior is still to be determined.

The agent term seems to be used in a different way from journals specialized in agents, such as Journal of Agent Oriented Software Engineering or Journal of Autonomous Agents and Multi-Agent Systems. The term agent is used as an equivalent of actor, like an abstraction of an entity that does things in an intelligent but undetermined way. There are works, like Farhangi [1], and Amin and Wollenberg [5], supporting Agents, but they do not tell or point out what an agent is for them. Moreover, they do not contain references to published works in conferences or journals specialized in this topic.

The IEEE Power and Energy Society MASWG studied this problem too. They issued two reports which were written as summaries of existing agent research results rather than guidelines for the application of Multi-Agent Systems (MAS) to the Power Grid. The first one [9] justified the interest of agent technology for Power Grids, concluding that a MAS can be used either as a way of building robust and flexible hardware/software systems or as a modeling approach. Also, the report indicated intelligence is realized by the incorporation of known AI techniques. MASWG point at areas where MAS can contribute such as distributed control, modeling and simulation (e.g., energy markets), protection, and monitoring and diagnosis. The second report [10] is more relevant to this paper since it deals with techniques and tools that could allow engineers to use MAS. The second report starts with an analysis of the FIPA (Foundation for Intelligent Physical Agents) standards for communication and interoperability. It puts special emphasis on using an ontology for Power Grids, more precisely the one previously mentioned. In theory, this ontology could facilitate the understanding of different MAS in a Power Grid. The design of the MAS, which is the focus of our contribution, has also a dedicated section, but smaller than the previous ones. The authors in [10], do not provide detailed recommendations but references to some AOSE methodologies. The also use some words about agent architectures (agent anatomies in the paper) and recognize JADE [11] as the main agent platform implementation. The report also says that there is insufficient information to support a recommendation of any particular agent anatomy. Furthermore, MASWG seems not to have had any activity since 2011, so no additional research could be found on this topic. Anyway, the results provided by the two reports, in our opinion, are not clear guidelines or recommendations of how to apply MAS in a Power Grid. For instance, blueprints of specific and recommendable configurations of MAS for Power Grids are missing. However we consider that the two reports are a very relevant starting point. Our contribution could be considered as a continuation of this effort since we show how we apply AOSE solutions to a particular problem and discuss the differences regarding the original version.

Some of the participants of the MASWG, later on, issued a new report [12] that focused on the performance of existing agent based solutions for Power Grid control. They briefly refer to AI techniques for power flow management, AI techniques for voltage control and AI techniques for restoration. They also consider how intelligent systems can be used to realize a Smart Grid architecture. More precisely, they consider distributed and robust deployment but also Self-Management through condition monitoring and selective devolved control. We share these concerns too, but again, we feel a critical analysis of current MAS solutions needs to be done in the context of Smart Grid. A system providing the features cited in papers like [12] are not necessarily made following trends in agent research or conventions in AOSE. We are not defending that the latter the better, but just that they are the result of a deep discussion on how agent features could be designed.

In the following lines we do a quick review of some works that propose the use of MAS as a solution, focusing on the problem of Microgrids management. As it can be observed, engineers are mostly using an agent platform, such as JADE, to develop a solution. As we discussed at the beginning of Section 2.1, using an agent platform does not mean it is being used properly. Just as if a programmer uses an object oriented language does not imply object oriented programming is being made. The following cases are examples of this argument.

Pipattanasomporn *et al.* [4] is a highly cited work which contains meaningful references to agent research and tries to translate those concepts to Microgrids, focusing on the energy distribution level. They propose Intelligent Distributed Autonomous Power Systems (IDAPS) that pursues to control a blackout. It is developed using ZEUS [13] which has graphical environment that facilitates the creation of such systems. The Microgrid is modeled in Matlab/Simulink and connects the agents as actuators. The intelligence, from an external point of view, consists on a mechanism for transition from connected mode to island mode in case of a blackout. However, this paper does not regard the scalability of the solution and how agents do interact to solve problems. In fact the control is embedded in a single agent which is the one that takes the decisions.

Xiao *et al.* [14] also suggests to combine Matlab and ZEUS to realize a MAS controlled Microgrid. They propose a hierarchical control system to replace a SCADA (Supervisory Control And Data Acquisition) based system, both are frequently centralized solutions. The hierarchy that Xiao *et al.* defend creates different levels of responsibilities: a distribution master station level, a distribution substation level and a distribution terminal level one. Authors follow the ZEUS steps in a more explicit way than [4] and deploy the system in a distributed way across several computers. However, the solution resembles to Pipattanasomporn *et al.* [4] at the lower level. There is a Distributed Energy Resource (DER) agent and a Load agent and Substation agent. They play similar roles than the [4] approach. There is also a coordination protocol among levels but it is not detailed. Moreover, some designing artifacts of the solution are missing in the proposal. We find again that the expected intelligence of the system is not discussed. The experiment implements a straight deployment with few agents. Unfortunately, there is short information that allows to reproduce the experiment.

Oyarzabal *et al.* [15] addresses a Microgrid management system built using agent technologies. Authors use JADE platform to develop the system, but they are not using a disciplined approach. They identified up to ten agent types being each of them responsible of a specific part of the system. This is a design that seems to follow a familiar organization of roles in the Microgrid. Unlike others, here there seems to be a concern for identifying different agent types but it does not address a behavior beyond as specific protocol implementation. Each agent implements part of one algorithm for the control of the system but there is no concern for the role of intelligence in this kind of approaches. The work succeeds in the experimentation to prove that a JADE based implementation of such systems could work at least in terms of scalability. However in this solution agents do represent mainly a division of labor. The benefits of agent technology in this case are similar to those ones of other distributed programming technologies. CORBA (Common Object Request Broker Architecture), for instance, could have been used to distribute responsibilities in a similar way.

Jiang [16] introduces an agent based control framework for DER Microgrids. One of the reasons for using agents is that these kind of Microgrids are in favor of energy sources distribution, and therefore its control should be distributed as well. The author suggests three kind of agents: an agent representing the energy source, a second one that represents the energy storage and a third one to represent a load. In this case, no agent oriented technology is cited or how the inherent features of the agents are relevant in the experiment. This kind of contribution is frequent; for instance we can find them in agent conferences, such as the "International Conference on Autonomous Agents & Multiagent Systems" (AAMAS), which are not the most relevant ones.

However, it is possible to use agents just as a convenient concept for thinking about the problem. Afterwards, this concept can be realized through agents or any other alternative technology. This is the case of Ramchurn *et al.*[17] that studies a strategy for reducing peaks of demand on behalf of consumers by using agents. However, the authors do not cite the technology used to execute the experiments. Therefore, it is hard to evaluate the relevance of agent research in the experiment itself. We also found a short design in these approaches, but the way the agent technology is used cannot be evaluated properly.

Kouluri and Pandey [18] focus on Microgrid control using agents too. JADE is the chosen agent platform. The authors identify three kind of agents: the control agent, the DER agent and the load agent. There is also an flow chart representing the activities in the system. These activities are coded within the abovementioned agents. We assume that this flow chart summarizes the expected intelligence of the system. In this case, this chart permits to issue the appropriate orders to each agent in order to handle the blackout. In the experiments, agents are connected to a Simulink instance, which realizes the Microgrid simulation. Simulink provides the input to the agents and also executes the orders generated by them. Like in [4] authors discuss the ways that the system can make the transition from island to connected mode. Despite the interest of the contribution, there are no details of specific agent features. Like other similar papers, we think that this kind of experiment proves the middleware capabilities of JADE, mainly. Other more inherent agent features, such as peer-to-peer communication, the use of ontologies, autonomy, or mobility, are missing.

Dimeas and Hatziargyriou [19] present a MAS for the control of a Microgrid which is an evolution of a previous work [20] and it is implemented in JADE. The proposal distinguishes between three levels: the distribution network operator (DNO) and market operator (MO), at the level of medium voltage and coordinating different Microgrids; the Microgrid central controller (MGCC), which acts as mediator between the Microgrid and the network/market; and the Local Controllers (LC), which controls DERs, production and storaget units, and some local loads. The market side part, influences the production in a Microgrid taking into account that energy could be bought or sold. There are roles identified for agents with specific responsibilities. However, within a Microgrid, there will be only one MGCC and several LCs. In the system, there will be only one DNO, MO, but several MGCC (one per simulated Microgrid). Authors compare a previous solution [20], where control in the Microgrid was centralized in the MGCC, against another that delegates control to the LCs and coordinates them through auctions at the market level [19]. Authors use auctions as main mechanisms to decide what to do, *i.e.*, the main mechanism to produce the desired intelligent behavior. Authors also use ontologies to facilitate inter-agent communication, though its benefits are not explored in detail. The coordination issues between the Microgrid and the market are studied, too, though there is no formal solution for this issue, only ideas of how to deal with it.

Kumar Nunna and Doolla [21] propose a MAS architecture for modeling the electricity market and implements a prototype using JADE. Agents are distributed in two levels: field level, to manage individual micro-grids to match the supply and demand, and market level, to cater the trading among the Microgrids and with the grid. The authors propose the matching of production and the demand through auctions implemented as FIPA auction protocols over the JADE platform. Simulated Microgrids are capable of storing/releasing electricity to the grid. There is a scheme of the relevant agents and, unlike others, authors show test cases involving more than one Microgrid. The solution is not completely decentralized since there is only one market, with its corresponding agents, that launches the auctions. In fact, it is this market the one that coordinates the bids and decide what to be produced/consumed.

In [22], a decentralized control architecture for Microgrids is presented. Even though the authors describe the system as a MAS, its architecture is not introduced. Authors introduce two scenarios where the Microgrid has to adapt to new conditions, like loss of solar resources. JADE platform is used to implement the communication interface between agents and the Microgrid, which is simulated with MATLAB/Simulink. The expected behavior is a complex one, close to what one may consider "intelligent". According to the text, the complexity arises because of the the variations of prices (both sell and buy) of energy in the utility market. This variation enforces that a Microgrid, sometimes, prefer to sell energy or to buy it. Also, the solution is a self-organizing one where there are negotiations at several levels. The proposed solution is based on a series of individual utility functions that realize the decision making process. These ideas seem close to the results agent research pursues. However, no system blueprints are provided. The benefits of having used JADE are not discussed either.

Logenthiran *et al.* [23] present a MAS for real-time operation of a Microgrid. The multiagent system consists of several distributed generator agents, load agents, a renewable energy source agent, a storage system agent, a microgrid manager agent, a schedule coordinator agent (SC Agent), a demand side management agent (DSM Agent) and other administrative agents. Comparing to other approaches, SC Agent and DSM Agent can be highlighted. The SC Agent executes a two-stage scheduling: day-ahead and real-time scheduling while the DSM Agent performs load shifting before the day-ahead scheduling and reduced the load in real-time whenever it is necessary and possible. As in many other works, a design of the MAS is not presented although JADE platform is used to implement the system. The number of instances of each agent is limited as well. Load agents are expected to be found in undetermined numbers. However, the paper only mentions one SC Agent.

In conclusion, we can observe that the cited works utilize an agent platform implementation to develop their MAS, being JADE the most used (see [15,18,19,21–24]). However, we consider that the reviewed works do not make an extensive use of the key features of agent technology in the context of Microgrids:

- In most cases, there is one instance of each agent. Hence, authors do not consider any interaction between similar agents. As we will show later, the possibility of having more than one instance of an agent may affect positively the performance of the system in the case of multi-agent systems [25].
- Agents do have little choice or the choices are not evident in the approach. Many of the cited papers consider the autonomy of the agent as a relevant feature. However authors are not making explicit the alternatives that each agent has got or even considering that an agent refuses to produce electricity, for example.
- Decentralization is done through distribution. There is a control centre which can be allocated in different places, but there is just one control element. Decentralization happens when the control is

distributed along the different existing agents. A trivial case would be having more than one control element. The reviewed literature is mostly centralized eventhough proposed MAS are distributed. There is a tendency towards having a distinguished agent that controls or coordinates everything.

- Agent to agent relationship is often a client-server one. The Peer to peer approach, which is more natural to agent technology, is not present in the reviewed papers. By using JADE, any agent can address any other agent. However in the reviewed papers agents can address other specific agents and only in a very particular way . Furthermore, in the selected papers, agents play just one role. This reduces the flexibility that could allow a purer MAS solution.

- There are no blueprints. What authors introduce as design is usually a graphical depiction (custom made) of key elements in the system together with annotations. These graphics are not useful for detailing an agent based system and there are already methodologies for capturing and specifying MAS since, at least, 2002 [26].

## 2.1. Reflections on the Use of Agents

The transition to new technologies implies having discussions about what approach works better. It happened with the transition from unstructured to structured programming (which eliminated the GOTOs statements among others) and the transition from Modules to Objects. The latter is particularly relevant to this problem because it leads to a book [27] about how Object Oriented Programming was wrongly used. It could be considered that this book inspired Wooldridge and Jennings [28] to apply the same principles to the world of agents. Wooldridge and Jennings identified some frequent approaches which they considered bad practices or, at least, issues to consider. Among them, there are the following stances:

- Believing that agents can solve any problem;
- There are too many agents or not enough ones in the solution;
- Agents are oversold: "the design is good just because it has agents";

- Becoming dogmatic about agents or an agent evangelist: "agents can do anything!";
- Designers do not know why they are using agents. In fact, other technology may suit better the problem.

The first and the last points are specially significant because of how easily is to fall in one of them. They are not technological mistakes but a result of the attitude of the developer. As a guideline for developments and to prevent falling into one of those traps, the key could be to use software engineering. Gómez-Sanz [29] recommends three basic principles:

- Agents are programs at the knowledge level. Newell [30] say that "an agent is a program that processes knowledge". Moreover an agent satisfies the rationality principle. This principle states an agent which chooses those tasks that allow the Agent to reach its goals. Choosing goals, finding ways to achieve them and determining how the agent can create new knowledge are all the elements that, together, make an agent something different.
- An agent is a container for Artificial Intelligence solutions. The main text book in Artificial Intelligence, Russel and Norvig's [31], presents agents as the wrapper for algorithms created within classic AI. The kind of intelligent behavior and the techniques to achieve it can be obtained in the AI literature. This is useful when a single agent is the one to show intelligence. However, when the combination of individual behaviors is the goal, classic AI is less useful. Distributed AI is another line and can be regarded as the beginning of agent research. Relevant agent based systems have rarely only one agent, they usually have many.
- The Hunhs and Singh's [25] test assumes that adding a new agent to an agent oriented system will alter its global behavior in some way. The improvement of the system performance depends on the agent and its purpose. The test precisely proposes to replicate instances of the existing agents trying to find these improvements. It also suggests having an utility function to be defined before starting the test in order to measure this fact. When there is none may be due to the fact that the agents are not aware of the existence of others or that any benefits are reached from interacting with them. In some way it means Agents are not behaving as they should.

# 3. THE DISCIPLINE OF THE DEVELOPMENT OF MULTI-AGENT SYSTEMS

Agent Oriented Software Engineering (AOSE) intends to create a development discipline and normalize the use of software Agents. AOSE is gradually incorporating Model Driven Development practices. This implies that the developer must specify the problem using some modeling language and then conceive how this model is transformed either into other models or into executable code. As a result, current AOSE methodologies can produce full MAS and/or provide an advanced starting point for the development.

However, the use of AOSE methodologies does not guarantee that bad habits disappear. Many of the issues pointed out in [28] (see Section 2.1) are caused by the attitude of the system designers rather than the technology itself.

INGENIAS [6] is an Agent Oriented Methodology and it is supported by the INGENIAS Development Kit [32]. It produces JADE [11] Agents and it is the most widespread platform for agent development. Hence, it is possible to take advantage of existing software developed for JADE. INGENIAS has its own notation for representing MAS (see Figure 1). INGENIAS can switch or mix this notation and UML notation with stereotypes. The latter is frequently used in documents for people with previous experience in this methodology.

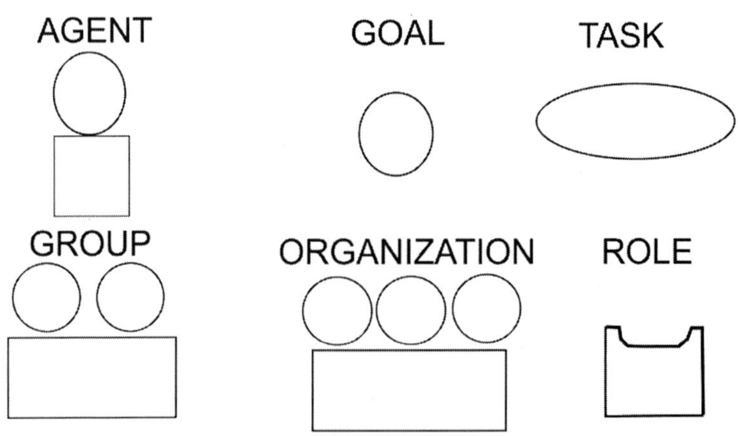


**Figure 1.** Meaning of some icons used in the system specification with INGENIAS.

In this paper, INGENIAS is applied to a case obtained from a frequently cited conference paper introducing IDAPS [4]. The refactored version will be named INGENIAS IDAPS or I2DAPS. It illustrates a few modifications to the original work that show a different, more agent-like, approach to get similar results to the original work. It is not our purpose to show I2DAPS is better than IDAPS. It is an example of what could have been done if ideas from Section 2.1 had been used.

There were four agents per Microgrid in the original work:

- User Agent. A gateway for interacting with the user.
- DER Agent. It registers a control agent, communicates with other DER agents, senses and reacts to measurements, controls the associated DER, informs of power production and receives requests from the user agent.
- Control Agent. It registers elements from the Microgrid, monitors the system, analyses the situation in the system, stores data about the system, sends signals to the main circuit breaker when grid failures are detected and provides the user a graphical representation.
- Database (DB) Agent. It registers information that is produced in the system.

IDAPS used ZEUS [13] to create the system. ZEUS was a reputed WYSIWYG ("What You See Is What You Get") editor for MAS that was discontinued many years ago. Anyway its quality is still outstanding. Therefore it could be said that IDAPS was built using solid agent oriented tools. ZEUS was also the name of a body of recommendations on how to use the tool to build systems. IDAPS paper does not elaborate on the application of ZEUS but there are some decisions which were not justified such as stating a DB should be seen as an agent. ZEUS also underlines the importance of defining the deployment but this information was not included in the paper beyond citing there was one instance of each agent type. Hence, there are two elements to be evaluated that could lead to important differences: the way the agents are used and the way the results are presented.

The refactored I2DAPS is introduced first in Figure 2.

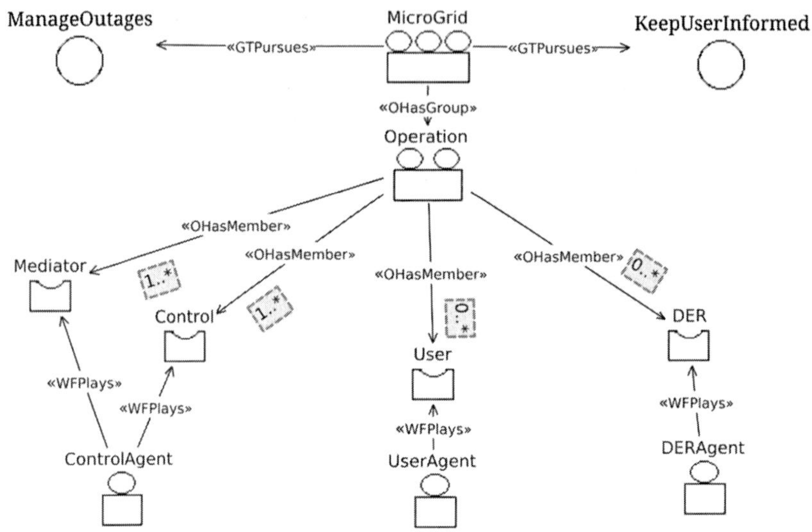


**Figure 2.** Agents of I2DAPS associated to an organization. (DER means "Distributed Energy Resource").

It uses INGENIAS notation and identifies the roles of the agents in the system together with the context of these roles. The most meaningful changes are:

- In IDAPS, there was no first class representation for the Microgrid. Agents were not declared as belonging to a Microgrid either. They just existed and it was the actual programming what made them aware there were other Agents belonging to the same Microgrid. By associating these Agents to the group and organization, this issue becomes transparent for the reader. The organization will be the Microgrid and the distinct activities involved into its operation left to specialized groups.

- We have added an Operation group to distinguish between the different business and management processes that a Microgrid requires. Other groups may realize other aspects, like Billing.

- Roles have been included in the specification. ZEUS methodology does include the concept of role but the paper describing IDAPS is omitting that part. A role is an aggregation of tasks, goals and information that can be assigned to different Agents so that we achieve an important degree of reuse. In our refactored solution, we distinguish between Control, User and DER roles. Agents will play these roles: all of them, some or just one. This gives the designer more freedom for defining what kind of Agents will exist. There is also one more role, the Mediator. This role was included to achieve a more peer-to-peer solution and represents the agent that promotes the obtaining of consensus as it will be introduced in the next section.

- The DB Agent of the original paper has been eliminated. It did not seem to be a relevant agent if the definition from Section 2.1 was used. A DB does not need to satisfy the rationality principle; in INGENIAS, for instance, the DB fits better as a resource used by some other existing agents or associated to any group. If it is associated to the group, all agents belonging to it will have access to the DB.
- There are explicit goals. The new system defines goals the system must achieve. The first goal is to keep the user informed of what is happening inside the Microgrid. The second goal is to manage blackout and electricity outages. System actions and roles are related to these two goals, so their purpose is explicit.

These ideas are presented in Figures 2 and 3. Figure 2 represents the Microgrid with the organization Microgrid. Being an organization, it is granted the entities can subscribe/unsubscribe from it. Moreover the diagram establishes that the Operation group must be formed by exactly one agent playing the role Control, any number of Agents playing the role of User and any Agents playing the role of DER. For each role, the same diagram identifies three different Agents. In this case, we assume that the Control Agent is playing two roles. This enables this agent to act as a mediator or as a controller. This is another example of flexible design which was not found in papers from Section 2.

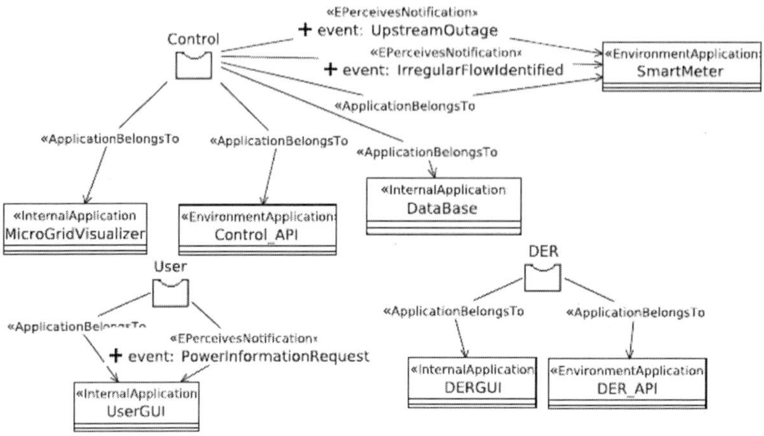


**Figure 3.** Resources associated to system roles. (API: "Application Programming Interface"; and GUI: "Graphical User Interface".)

Figure 3 explains the resource allocation per role, something that at the IDAPS paper was made in plain text. These resources are grid control libraries (e.g., for activating circuitbreakers), DER libraries (for accessing DER infrastructure), visualizers for the user, meters and some GUIs (Graphical User Interfaces). The previous elements may provoke some interesting situations in the grid that should be controlled such as the existence of an upstream outage or the detection of a low quality of the electricity flow.

As a key difference from works cited in Section 2, there is no need to represent as Agents something that it is not. This is the case of the above mentioned DB, but there are others as the way the agent interacts with the DER or any other hardware element.

Readers can find links between this solution and the hierarchical approach from [14]. The definition of groups helps to define controlled scopes for the agents to interact. The use of these groups will be more evident in next section where a deployment with multiple groups is used.

IDAPS proposes three basic functions: registration/deregistration of devices into the Microgrid, managing blackouts and informing the user on DER status.

The first is a function already provided by the organization in INGENIAS. IDAPS is not dealing with the situation whether the registration implies something else besides annotating there is one more element. This information may be useful elsewhere, for instance, to take later decisions. In any case, organizations in INGENIAS already provide subscription mechanisms and the organization information can be used to limit the scope of agent interactions (see Section 4).

The second is represented in Figure 4. According to Pipattanasomporn [4], there is a single agent in charge of managing the blackouts through the clever use of circuitbreakers. We translate it as a single task that, given a blackout event detected by low level hardware, it proceeds to create an island and isolate from the main lines. The triggering event is the upstream outage that is produced by the applications from Figure 3, in particular, by the SmartMeter. The task assigned to the control role uses the control API (library) to enable the circuitbreakers. The ManagedDeviceID element informs the tasks which device is assigned to this agent for its management. Each agent can be assigned to different devices so that we need to make explicit this assignment. In addition, it is made explicit that this task contributes to the achievement of the goal Manage outages, already identified in Figure 2.

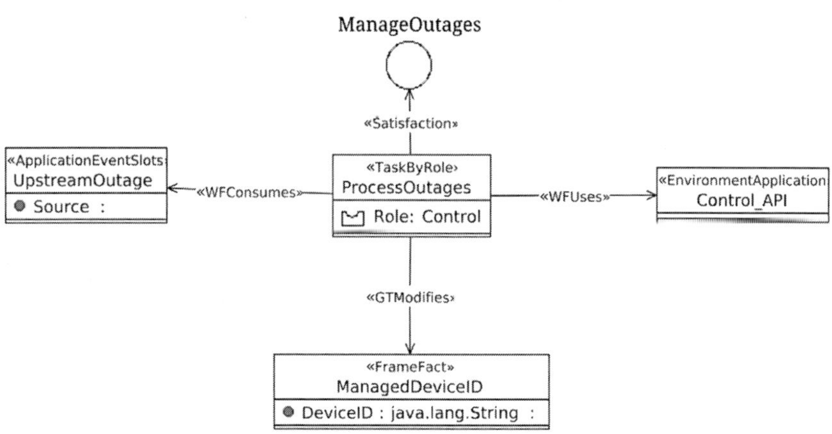


**Figure 4.** Blackout management task.

The third function implies the collaboration of several agents to inform the user of the current status of the grid. This is necessary to attain the goal keep the user informed identified in Figure 2. For this issue we use an INGENIAS workflow definition in Figure 5. The notation includes interconnection of tasks within the workflow pointing which element is produced by one task and which is consumed by the next task in the sequence. The "+/−" symbols mean that the referred piece of information is created, "+", by the previous task and consumed, "−", by the next task. Roles responsible of each task are declared together with the tasks.

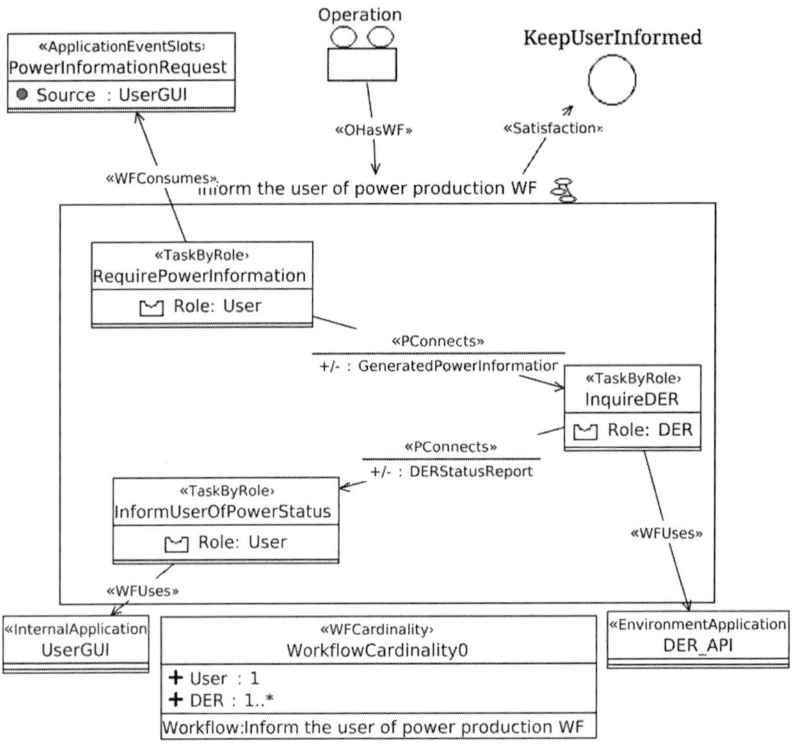


**Figure 5.** Workflow (WF) for informing the user on the current status.

The workflow is initiated by the user for requesting more information (see task require power information andFigure 3 to identify the source of this event). This information is obtained from the DER (see task inquire DER). Once it is collected, the information is presented to the user through a GUI (see task inform user of power status). Furthermore the workflow has as scope the operation group type (see Figure 2). Since there can be several group types in the same Microgrid, this constrains the communication to agents belonging to the same group. The number of agents involved in this workflow is not fully determined until runtime.

There is one user involved and at least one DER (see entity workflow cardinality). The default behavior implies that one user will ask all DERs within the same Operation group. The workflow pursues the goal keep user informed too. This goal is also associated by default to the different tasks that belong to this workflow.

The deployment of the Microgrid was not considered in Pipattanasomporn [4]. Figure 6 shows how it would be done in the case of INGENIAS. The central element is a Deployment Package which represents an executable that launches a specific set of agent instances. The deployment represents the intended structure of the organization. In this case, there is one operation group with one Control Agent, two DER Agents and three User Agents.

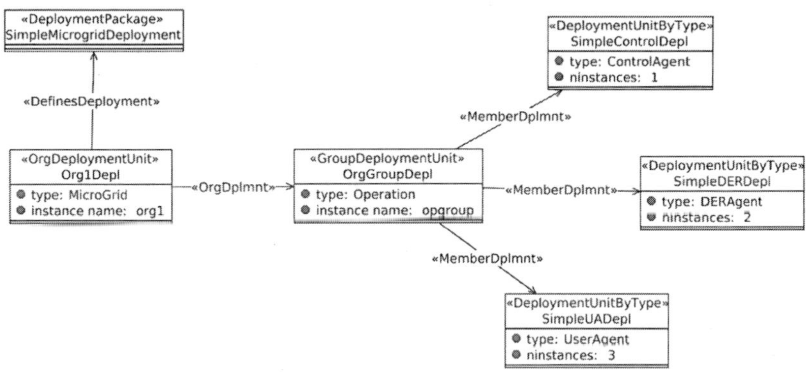


**Figure 6.** Initial deployment description of the system.

Once designed, in INGENIAS diagrams are automatically processed and fully functional agent systems are produced. The generated system already implements the designed tasks and is capable of executing the expected protocols as shown in Figure 7. The development is managed using the Maven tool [33] which allows, with a minimum installation, to address the development in a efficient way.

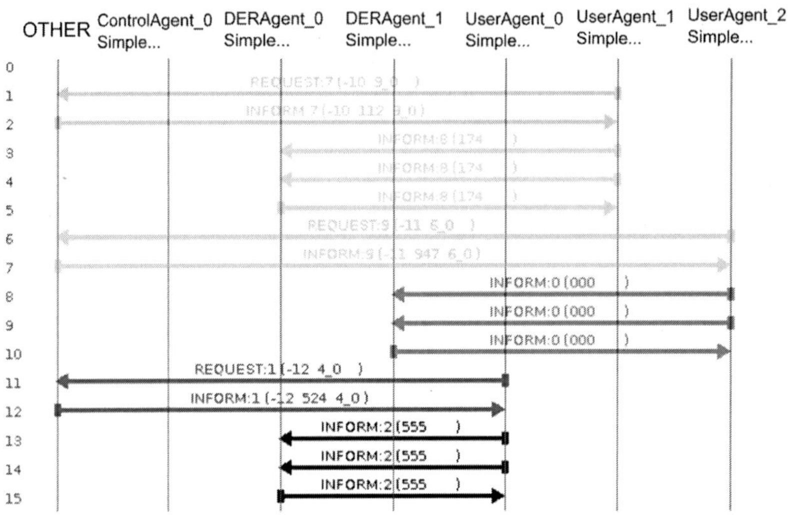


**Figure 7.** JADE Sniffer showing how the communication among agents is taking place.

The execution and message interchange among generated agents is represented in Figure 7. Produced agents are JADE ones where JADE platform is the most widely used. Moreover, generated agents can reuse JADE tools like the sniffer. The sniffer intercepts all messages interchanged among agents and presents sequence diagrams such as those of Figure 7. The numbers on the left indicate the sequence order; the columns, the different senders/receivers; and the labels over the arrows, the message type. This initial version can be downloaded from the site [34] and github [35]. The figure shows how each user agent is querying DER agents in the system to obtain a status report, like in the original work. Looking at the message number, on the left, queries are messages 3, 4, 8, 9, 13 and 14; and answers are messages 5, 10 and 15. The apparent duplication of queries is due to the way inter-agent conversations are automatically coded in INGENIAS.

### 3.1. Evaluating I2DAPS

So far, a humble translation of Pipattanasomporn's work [4] has been introduced. The straight application of INGENIAS can bring some benefits with respect to the original IDAPS work which will be detailed below:

- This representation of the agent concept is more evident when introducing the deployment (see Figure 6). Reminding the principles introduced in Section 2.1, INGENIAS forces the developer to think how many instances of each agent should exist. In the next section, this example will go one step further and detail how each instance can be specified.

- There are blueprints. The representation of the agent elements in the system and their interdependencies are more explicit, which facilitates explaining others what the system does. For each agent, there is an account of the capabilities of each of them. Each capability is connected to a system goal. Most of the concepts in INGENIAS are connected to goals in one way or another. Since developers identify system goals, it is easy to justify the convenience of one element with respect the original specification requirement. It is a matter of finding how the element, like a task, is connected to the system goal, which is a form of requirement traceability along the design. Also, this brief analysis has led to changes in the conception of the system. Elements like a DB have not been considered as agents.

- Agents in the system can play more than one role. Figure 2 contains an agent playing two roles. This enables the agent to participate more in the system. In addition, it increases its capabilities and, potentially, the flexibility of the system. After all, the agent can play several roles at the same time. As it will be seen in the next section, this leads to situations where an agent can try to interact with itself because it is playing both the roles of client and server at the same time.

- The decoupling between what an agent is and what is not is clearer in INGENIAS than in the reviewed works from Section 2. The concept of application serves well to indicate how agents interact with non-agentified elements. Furthermore it prevents the overuse of agents when they are not required, for instance the case of converting a DB into an agent.

However, the example of I2DAPS has a serious flaw: it is disconnected from a Microgrid. Despite this fact, the example illustrates the potential of the approach and that initial assumptions made in Pipattanasomporn [4] could have been different if an Agent Oriented Methodology, such as INGENIAS, had been applied.

Despite avoiding pitfalls cited in Section 2.1, it is not clear if the system, even refactored, is showing an intelligent behavior. The most promising part is the one dedicated to avoiding blackouts propagation. Though it is an important feature it could be discussed if, for an external observer, to avoid a blackout is a sign of intelligence.

# 4. INTEGRATING MICROGRID SIMULATION INTO I2DAPS

To address the issues raised in Section 3.1, this extension works on the agent principles from Section 2.1. It explores the effect of replication of single agents such as DER Agents and Control Agents. According to Hunhs and Singh's [25] test, the incorporation of new agents of the same kind should lead to an improvement of the functionality. Note that some replication was made in the previous section with little apparent impact.

In this section we focus on Control Agents. Replicating them means that the system becomes more tolerant to system failures. However, it is not possible to add more control elements lightly and therefore this section will try to provide a preliminary solution to the mentioned problem.

Another improvement over the previous approach is the connection of the system to a simulator, just as the original work. Pipattanasomporn used Matlab while the second one will be based on GRIDLAB-D [36], a power grid simulator that also is able to simulate Microgrids. The simulator had to be modified to act as a real time simulator capable to receive orders in runtime and a new Java layer was incorporated. These improvements facilitated a greater connectivity to the INGENIAS system. The Microgrid to which agents will connect is introduced in Section 4.1. The design variations that will allow interconnectivity are presented in Section 4.2.

## 4.1. Case Study Microgrid

For experimental purposes, a radial Microgrid with distributed generation and batteries has been designed. Its structure is shown in Figure 8. The grid is made of six transformation centres ($TC_1 ... TC_6$) connected in medium voltage (15 kV) to a central one ($TC_0$); the latter receives the energy in medium voltage from the substation. A transformer of 1000 kV A at the entry of the substation receives the energy in high voltage (45 kV) from an external supplier and transforms it to a medium one. Furthermore, each $TC_i$ represents a transformation centre where the energy is converted and driven from medium to low voltage lines and viceversa (maximum kV A supported by transformer and line distances are also shown in the figure).

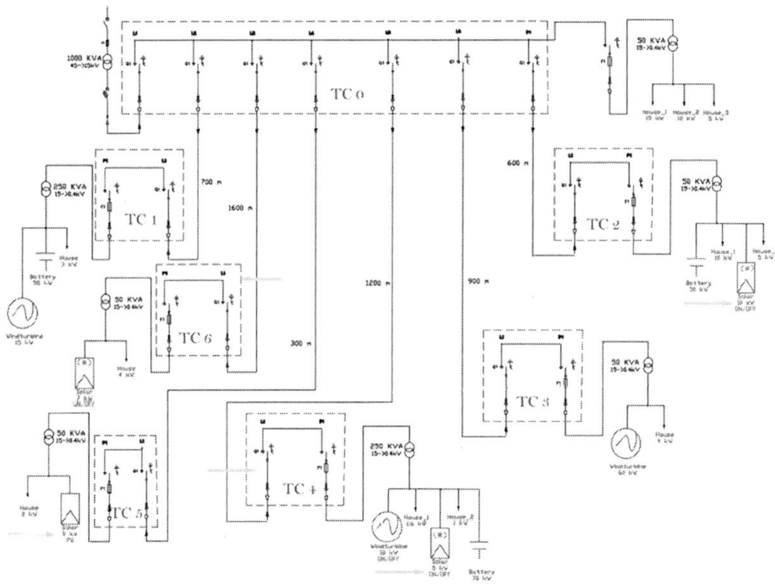


**Figure 8.** Microgrid to be controlled by I2DAPS.

On the other hand, the Microgrid has several elements all connected to low voltage lines such as energy suppliers (wind and solar generators), batteries and consumers. Power consumption or generation is specified in each consumer and supplier. In case of batteries, maximum charging/discharging power (kW) is provided. Note that the generators can be automatically controlled by the agent system. Each generator specifies its control mode which can be ON/OFF or full control of P and Q.

The solar generators of $TC_2$, $TC_4$ and $TC_6$ are not permanently connected to the grid so that they would work as plug-in elements. The connections of those elements with the Microgrid is represented with "(*)" in Figure 8. Moreover, the transformation centres or elements where agents have been connected are signalized with a green arrow.

For an easier comprehension of the grid, its final elements are described in the following Table 1.

**Table 1.** Elements specification.

| $TC_i$ | Transformer | Name | Control | Consumption (kW) | Generation (kW) | Fase | Distance to transf. (m) |
|---|---|---|---|---|---|---|---|
| $TC_0$ | 50 kV A | House_1 | - | 15 | - | ABC | 222.5 |
| | | House_2 | - | 10 | - | ABC | 498.1 |
| | | House_3 | - | 1 | - | ABC | 424.5 |
| $TC_1$ | 250 kV A | Windturbine | - | - | 15 | ABC | 6 |
| | | Battery | PQ | 50 | 50 | ABC | 20 |
| | | House | - | 3 | - | ABC | 424.5 |
| $TC_2$ | 50 kV A | Solar | ON/OFF | - | 10 | A | 18 |
| | | Battery | PQ | 30 | 30 | ABC | 19 |
| | | House_1 | - | 10 | - | ABC | 42.3 |
| | | House_2 | - | 5 | - | ABC | 51.5 |
| $TC_3$ | 50 kV A | Windturbine | - | - | 60 | ABC | 6 |
| | | House | - | 4 | - | ABC | 110.5 |
| $TC_4$ | 250 kV A | Solar | ON/OFF | - | 5 | A | 80 |
| | | Windturbine | ON/OFF | - | 60 | ABC | 38 |
| | | Battery | PQ | 30 | 30 | ABC | 20 |
| | | House_1 | - | 26 | - | ABC | 61.5 |
| | | House_2 | - | 1 | - | ABC | 123 |
| $TC_5$ | 50 kV A | Solar | PQ | - | 8 | A | 34 |
| | | House | - | 2 | - | ABC | 231 |
| $TC_6$ | 50 kV A | Solar | ON/OFF | - | 7 | A | 25.3 |
| | | House | - | 4 | - | ABC | 145 |

## 4.2. Extending I2DAPS

The connection of agents to the Microgrid is done through the resources depicted in Figure 3. These resources are coded as mediators with the Microgrid using remote method invocation utilities. Then, the agents can deliver orders to the Microgrid in real time or obtain measurements from the sensors associated to it. This action implies few changes to the specification since it is only needed to add the new programming code to some specific elements.

In order to evaluate the role of the agents in the Microgrid from Section 4.1, the infrastructure introduced in the previous section is enhanced with new agent instances. The interactions among these new instances are analysed below.

Figure 9 introduces a new deployment where two groups are depicted whereas Figure 6 presented one. The main difference is that agents associated to one group will only see those ones from his same group if the developer wants it. Each group has a control agent instance and at least one DER agent. Group subgrid $TC_2$–$TC_4$have DER agents connected to $TC_2$ (Solar 10 point in Figure 8) and $TC_4$ (Solar 5 point in Figure 8). Group subgrid$TC_5$ and $TC_6$ has less elements. Control agents are attached to $TC_4$ and $TC_6$. As a difference with Figure 6, this time agents are individually initialized.

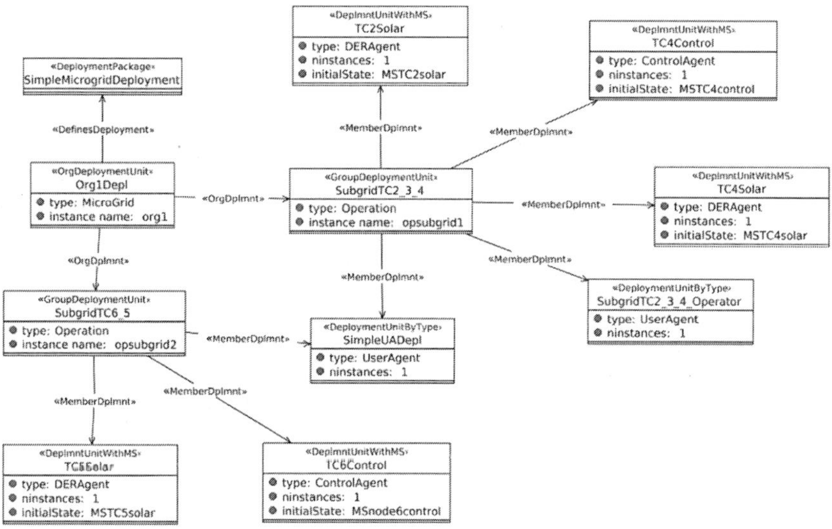


**Figure 9.** New I2DAPS deployment for the Microgrid from Figure 8.

The two groups determine two different scopes that will be more evident when explaining Figure 10. They limit the visibility of User and DER agents when performing certain operations.

Continuing the idea of replication of agents, Figure 11 does make an interesting point: what happens if there were more than one control agent in the grid? The system would become a chaos if the control is not properly partitioned. For this reason and to deal with the decentralization of control, we suggest a solution inspired in previous work [37]. The idea is to perform rounds of inquiries to the different controlling agents using a mediator role to reach an agreement.

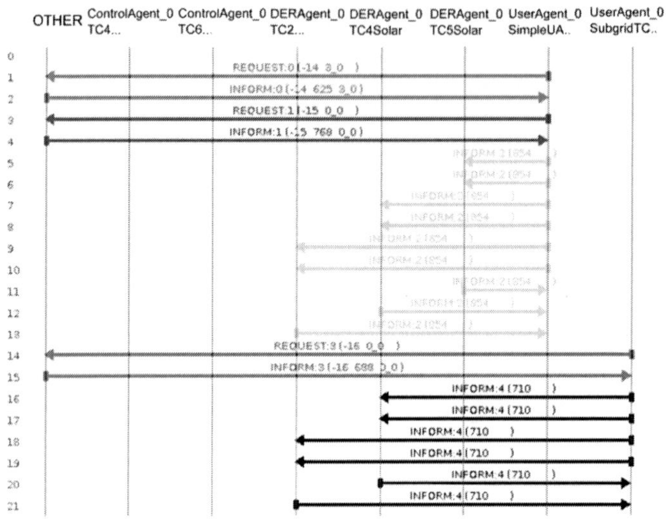


**Figure 10.** Interaction among User Agents and DER Agents limited by the group membership as described in Figure 9.

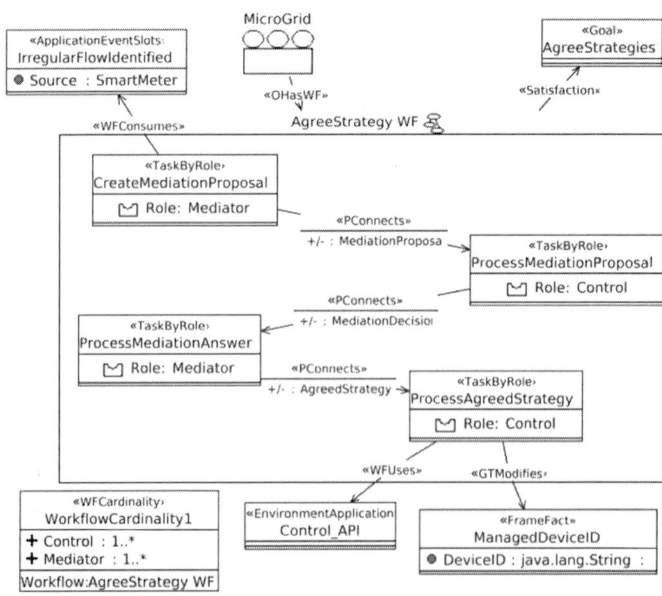


**Figure 11.** New workflow for arranging agreements among control agents.

This agreement workflow is showed in Figure 11. Control agents will now also play the role Mediator and Control at the same time (see Figure 2). Hence, given two control agents, both can start collaborating with each other with any combination of roles. This new workflow will permit control agents to interchange strategies and enact a chosen one. The mechanism is independent of the number of agents. It also will be sensible to the number of control agent instances, which ought not to collide and give conflicting orders to DER agents. Hence, any two control agents can try in any order to achieve an agreement acting as peers. This time the scope of the workflow is the organization so that two control agents, even if they are assigned to different operation groups, will be able to perform this workflow. The cardinality of the workflow (see entity workflow cardinality in Figure 11) is also 1 to *N*. There is one mediator but at least one control agent.

This sensibility is explicit in Figure 10, which is a more elaborated version of Figure 7. The figure shows how the scope of each user agent is different when contacting DER agents. One of the agents is belonging to two groups so it can retrieve descriptions of DER agents belonging to both groups too. Since the other user agent has access to only one group, using properly the organizational concepts it is possible to limit the number of acquaintances of one agent.

In the implementation, the GRIDLAB-D based Microgrid simulation will be started into another process. DER agents, when starting, will plug to their corresponding element in the grid and request a POWER ON command to the element. As a result, the GRIDLAB-D will take into account the energy production of the DER. Figure 12shows the Microgrid measurements taken at the main substation of the Microgrid. When DER agents are created and connected to the Microgrid the substation demand drops. It is coherent since incorporated agents are energy providers. Nevertheless, there is an increase of the distribution and transformation losses.

The influence of control elements can be seen only when there is any alarm. In this case, the alarm is the existence of an Irregular Flow Identified event generated by a Smart Meter (see Figure 3). Figure 13 shows such interaction. The interaction is arranged with all existing Control Agents in the Microgrid. Involved agents are Control Agent for $TC_4$ and Control Agent for $TC_6$. The second is acting both as leader and as subordinate. It works as leader when contacting other control agents and gathering proposals to solve the current issue. It works as a subordinate too when telling other control agents which strategy has been finally decided. In this case, the leader has to notify itself, since it plays both roles of mediator and control. Playing two roles in the same conversation means the same agent can play as message deliverer and receiver. This can be observed in lines 5, 6 and 7 of the Figure 13. In those cases, there is an arrow pointing at itself. Whatever the case is, the intention along the scenario is switching on associated devices in $TC_6$ and $TC_4$. This capability of adding elements to conversations while they are added to the system is one of the inherent capabilities of the agent technology.

The result of the interaction is presented in Figure 14. The agreed strategy consists in keeping generators on and discharging available batteries. Once again, the demand from the substation is reduced and there are some important losses in the system. The latter may be the case of a poorly designed Power Grid with less than adequate transformation centres. We use such elements to justify the activation of DER resources closer to the demand, specially when DER are not photovoltaic or wind based.

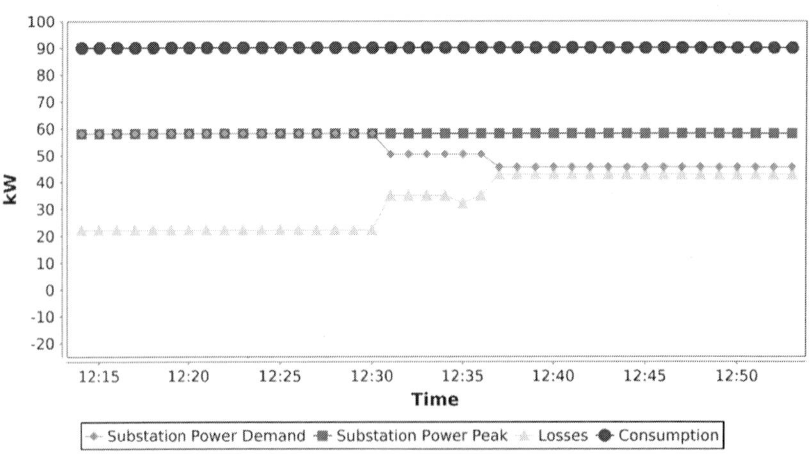


**Figure 12.** Measurements at the Substation when DER agents connect and enable the corresponding DER resources.

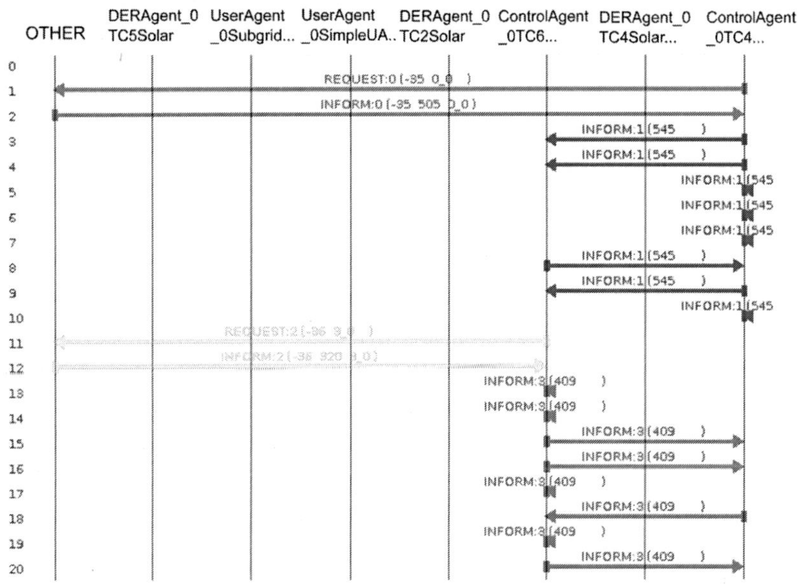


**Figure 13.** Workflow among Control Agents to deal with an extra demand.

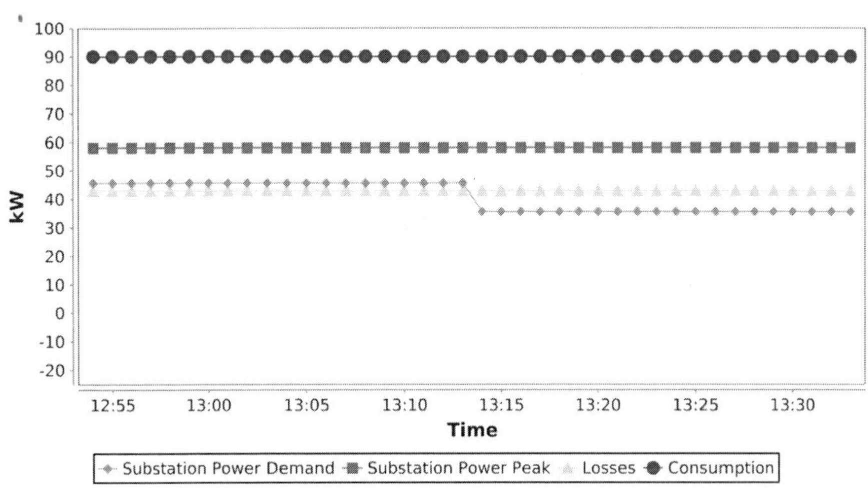


**Figure 14.** Demand from the substation when control is enabled.

## 4.3. Criticism of the Results

This new version is more complete than the previous one. Agents are connected to a power grid simulator and some charts could be elaborated with the results. Moreover, the visibility of the agents made possible to create distinct areas of influence while existing in the same Microgrid.

The intelligence of the resulting system is easier to observe than the last time but not absolutely complete yet. A real time operation of the simulated Microgrid may help for this task. If, during a long simulation, the resulting Microgrid behaviour showed improvements against the just turn everything on strategy, it would be a good evidence. Another hint could be to provoke failures in the network and check that the system is still capable of operating.

However, the agreement workflow (see Figure 13) allows the coexistence of different control agents in the same Microgrid, or even in the same group. As long as there is an agreement in the strategy and the willingness to comply with the agreements, the system should tolerate a number of control agents. In addition, this makes the system more fault tolerant. Control agents can fail and, still, the system could re-arrange remaining control agents and reassigning controlled devices if the agreement workflow were properly modified.

To conclude, thinking about the interaction among instances of the same agent was an educating experience. Agents can participate in the same conversation using different disguises, which can be an useful for combining capabilities in the same entity.

# 5. CONCLUSIONS

In this paper, we have reviewed some relevant research works about agents in Microgrids. Even though they apply agent technology, the resulting systems show some inconveniences: authors only consider one instance of each agent type; agents are not really having decision capabilities; there is distribution of components but little decentralization; agent interaction is a client-server one instead of peer-to-peer; and, most important one, there are no precise blueprints that explain what the MAS design is.

The role of agents in Smart Grids is not such evident for an agent researcher and, for this reason, we suggested in this work some guidelines to follow when designing an agent system to manage Microgrids. These guidelines are based on research trends pointed out in the agent research community.

In order to illustrate a practical example about how to carry out these recommendations, a specific highly cited work applying agent technology (IDAPS) to Microgrids [4] was evaluated and an alternative design was proposed (I2DAPS). The new design explores features of agent technology not addressed by the original authors. Among the changes performed over IDAPS, the most remarkable are:

1. Incorporation of the Organization concept to address the Microgrid subscription of agents. Also to illustrate how the different aspects of a Microgrid could be encapsulated within a single entity, the Microgrid Organization. Concretely, the paper includes the operation aspect, though not all power grid operation issues are considered.

2. Functionality reuse by means of roles. The original work simply assigned responsibilities to the agents whereas I2DAPS applies roles as mean to aggregate functions and associate them to different agents.

3. The not-everything-has-to-be-an-agent principle was applied and a clear distinction was made between what deserved to be considered an agent what did not. A DB in the original work was one agent and I2DAPS redefined it as non-agent element to be controlled by other agents.

4. Peer-to-peer philosophy was applied to realize cooperation between control elements. The example addressed several control elements which had to agree in a strategy to manage the network. The solution was an open one in the sense that new control agents could be incorporated to the agreement discussion.

5. In order to study the new system, agents were connected to a Power Grid simulator. Some charts with the tested Microgrid measurements and the dialogs among agents were included as well. Results showed how I2DAPS worked according to the previous description.

In addition, these benefits were represented in a way which is not usual in Microgrid papers that use agent technology. We applied INGENIAS to show how a blueprint of a MAS could be. INGENIAS and other agent oriented methodologies are solutions proposed by the agent research community to address an agent oriented development.

Even some benefits from the designed perspective have been identified in this work, there is still more experimentation to carry out such as simulating in a real environment or provoking different failures in the network in order to study how the system responds. Therefore, a performance study of the resulting solutions needs to be made. Though we show some experiments, more intensive ones are required to evaluate the impact of the new designs we propose. The performance of MAS systems has been already studied, but, as we justify in our study of existing works, we need to ensure before that the MAS technology is used as something more than a convenient middleware. An interesting experiment would be comparing IDAPS performance against I2DAPS, for instance.

## ACKNOWLEDGMENTS

To the project MIRED-CON IPT-2012-0611-120000, funded by the Ministry of Economy and Competitiveness. This project aims to bring new capabilities to conventional Microgrids in order to obtain improvements in billing and reducing energy distribution losses.

## REFERENCES

1. Farhangi, H. The path of the smart grid. *IEEE Power Energy Mag.* **2010**, *8*, 18–28.

2. Sobe, A.; Elmenreich, W. Smart Microgrids: Overview and Outlook. Proceedings of the "Smart Grid" Workshop of the ITG INFORMATIK 2012 Conference, Braunschweig, Germany, 16–21 September 2012.

3. Hassan, R.; Radman, G. Survey on Smart Grid. Proceedings of the IEEE SoutheastCon 2010 (SoutheastCon), Concord, NC, USA, 18–21 March 2010; pp. 210–213.

4. Pipattanasomporn, M.; Feroze, H.; Rahman, S. Multi-Agent Systems in a Distributed Smart Grid: Design and Implementation. Proceedings of the 2009 IEEE/PES Power Systems Conference

and Exposition, PSCE '09, Seattle, WA, USA, 15–18 March 2009; pp. 1–8.

5. Amin, S.; Wollenberg, B. Toward a smart grid: Power delivery for the 21st century. *IEEE Power Energy Mag.* **2005**, *3*, 34–41.

6. Pavón, J.; Gómez-Sanz, J. Agent Oriented Software Engineering with INGENIAS. In *Multi-Agent Systems and Applications III*; Marík, V., Pechoucek, M., Müller, J., Eds.; Springer: Berlin/Heidelberg, Germany, 2003; Volume 2691, pp. 394–403.

7. Brooks, R.A. Intelligence without Reason. In *Computers and Thought, IJCAI-91*; Mylopoulos, J., Reiter, R., Eds.; Morgan Kaufmann: Burlington, MA, USA, 1991; pp. 569–595.

8. Brooks, R.A. A robust layered control system for a mobile robot. *IEEE J. Rob. Autom.* **1986**, *2*, 14–23.

9. McArthur, S.; Davidson, E.; Catterson, V.; Dimeas, A.; Hatziargyriou, N.; Ponci, F.; Funabashi, T. Multi-agent systems for power engineering applications—Part I: Concepts, approaches, and technical challenges. *IEEE Trans. Power Syst.* **2007**, *22*, 1743–1752.

10. McArthur, S.; Davidson, E.; Catterson, V.; Dimeas, A.; Hatziargyriou, N.; Ponci, F.; Funabashi, T. Multi-agent systems for power engineering applications—Part II: Technologies, standards, and tools for building multi-agent systems. *IEEE Trans. Power Syst.* **2007**, *22*, 1753–1759.

11. Bellifemine, F.; Poggi, A.; Rimassa, G. JADE: A FIPA2000 Compliant Agent Development Environment. Proceedings of the Fifth International Conference on Autonomous Agents, Montreal, QC, Canada, 28 May–1 June 2001; pp. 216–217.

12. Davidson, E.; Catterson, V.; McArthur, S. The Role of Intelligent Systems in Delivering the Smart Grid. Proceedings of the 2010 IEEE Power and Energy Society General Meeting, Minneapolis, MN, USA, 25–29 July 2010; pp. 1–6.

13. Nwana, H.S.; Ndumu, D.T.; Lee, L.C.; Collis, J.C. ZEUS: A toolkit for building distributed multiagent systems. *Appl. Artif. Intell.* **1999**, *13*, 129–185.

14. Xiao, Z.; Li, T.; Huang, M.; Shi, J.; Yang, J.; Yu, J.; Wu, W. Hierarchical MAS based control strategy for microgrid. *Energies* **2010**, *3*, 1622–1638.

15. Oyarzabal, J.; Jimeno, J.; Ruela, J.; Engler, A.; Hardt, C. Agent Based Micro Grid Management System. Proceedings of the 2005 International Conference on Future Power Systems, Amsterdam, The Netherlands, 16–18 November 2005.

16. Jiang, Z. Agent-Based Control Framework for Distributed Energy Resources Microgrids. Proceedings of the 2006 IEEE/WIC/ACM International Conference on Intelligent Agent Technology, IAT '06, Hong Kong, China, 18–22 December 2006; pp. 646–652.

17. Ramchurn, S.D.; Vytelingum, P.; Rogers, A.; Jennings, N. Agent-based Control for Decentralised Demand Side Management in the Smart Grid. Proceedings of the 10th International Conference on Autonomous Agents and Multiagent Systems, International Foundation for Autonomous Agents and Multiagent Systems, AAMAS '11, Richland, SC, USA; 2011; Volume 1, pp. 5–12.

18. Kouluri, M.; Pandey, R. Intelligent Agent Based Micro Grid Control. Proceedings of the 2011 2nd International Conference on Intelligent Agent and Multi-Agent Systems (IAMA), Chennai, India, 7–9 September 2011; pp. 62–66.

19. Dimeas, A.; Hatziargyriou, N. Operation of a multiagent system for microgrid control. *IEEE Trans. Power Syst.* **2005**, *20*, 1447–1455.

20. Dimeas, A.; Hatziargyriou, N. A Multiagent System for Microgrids. Proceedings of the 2004 IEEE Power Engineering Society General Meeting, Denver, CO, USA, 6–10 June 2004; Volume 1, pp. 55–58.

21. Kumar Nunna, H.; Doolla, S. Multiagent-based distributed-energy-resource management for intelligent microgrids. *IEEE Trans. Ind. Electron.* **2013**, *60*, 1678–1687.

22. Colson, C.; Nehrir, M. Comprehensive real-time microgrid power management and control with distributed agents. *IEEE Trans. Smart Grid* **2013**, *4*, 617–627.

23. Logenthiran, T.; Srinivasan, D.; Khambadkone, A.; Aung, H.N. Multiagent system for real-time operation of a microgrid in real-time digital simulator. *IEEE Trans. Smart Grid* **2012**, *3*, 925–933.

24. Dimeas, A.; Hatziargyriou, N. Agent Based Control of Virtual Power Plants. Proceedings of the International Conference on Intelligent Systems Applications to Power Systems, ISAP 2007, Niigata, Japan, 5–8 November 2007; pp. 1–6.

25. Huhns, M.N.; Singh, M.P. Agents on the web: The agent test. *IEEE Internet Comput.* **1997**, *1*, 78–79. [Google Scholar]

26. Caire, G.; Coulier, W.; Garijo, F.J.; Gómez-Sanz, J.J.; Pavón, J.; Leal, F.; Chainho, P.; Kearney, P.E.; Stark, J.; Evans, R.; *et al.* Agent Oriented Analysis Using Message/UML. In *Agent-Oriented Software Engineering II*; Wooldridge, M.J., Weiß, G., Ciancarini, P., Eds.; Springer: Berlin/Heidelberg, Germany, 2001; Volume 2222, pp. 119–135.

27. Webster, B.F. *Pitfalls of Object-Oriented Development*; M & T Books: New York, NY, USA, 1995.

28. Wooldridge, M.; Jennings, N.R. Pitfalls of Agent-Oriented Development. Proceedings of the Second International Conference on Autonomous Agents, Minneapolis, MN, USA, 10–13 May 1998; pp. 385–391.

29. Go´mez-Sanz, J.J. The Construction of Multi-agent Systems as an Engineering Discipline. In *Engineering Societies in the Agents World VII*; O'Hare, G.M.P., Ricci, A., O'Grady, M.J., Dikenelli, O., Eds.; Springer: Berlin/Heidelberg, Germany, 2006; Volume 4457, pp. 25–37.

30. Newell, A. Reflections on the knowledge level. *Artif. Intell.* **1993**, 59, 31–38

31. Russell, S.; Norvig, P. *Artificial Intelligence: A Modern Approach*; Prentice Hall: Upper Saddle River, NJ, USA, 2010

32. INGENIAS Development Kit. Available online: http:// ingenias.sf.net (accessed on16 May 2014).

33. Maven-Apache Foundation. Available online: http:// maven.apache.org (accessed on 16 May 2014).

34. Agents for Energy-GRASIA Reseach Group. Available online: http:// grasia.fdi.ucm.es/energy (accessed on 16 May 2014).

35. GitHub. escalope/i2daps. Available online: https:// github.com/escalope/i2daps (accessed on 16 May 2014).

36. GridLAB-D Simulation Software. Available online: http://www.gridlabd.org/ (accessed on 16 May 2014).

37. García-Magariño, I.; Gómez-Sanz, J.J.; Pérez-Agüera, J.R. A Multi-Agent Based Implementation of a Delphi Process. Proceedings of the 7th International Joint Conference on Autonomous Agents and Multiagent Systems, AAMAS 2008, Estoril, Portugal, 12–16 May 2008; Volume 3, pp. 1543–1546.

# 8

# CHAPTER

# DUAL LAYERED ARCHITECTURE FOR MULTI AGENT BASED ISLANDING AND LOAD MANAGEMENT FOR MICROGRIDS

**Al Kulasekara[1], Ktmu Hemapala[2], Rarc Gopura[1]**

[1]*Department of Mechanical Engineering, University of Moratuwa, Katubedda, Sri Lanka*

[2]*Department of Electrical Engineering, University of Moratuwa, Katubedda, Sri Lanka*

## ABSTRACT

This paper proposes a novel dual layered multi agent system (MAS) based control system for the use in microgrid operations. In developing a smarter grid capable of withstanding disturbances and/or outages and providing quality service to the consumers, reliable microgrid control architecture is vital. The innovative microgrid control system proposed, makes the microgrid capable of isolating the local grid from effects of any upstream disturbances in the main utility grid by operating disconnected from the main utility via islanding, and it allows the most critical local loads to be supplied by any, available, local power source during such islanded operation. The proposed MAS control architecture is developed using the

JADE platform and it is used to control a test network simulated in MATLAB. The results of these simulations show the capability of developing MAS based reliable control mechanism for islanding and load management of microgrids based on the proposed concept.

**Keywords:** Autonomous Agents, Load Management, Microgrids, Multi-Agent Systems

# 1. INTRODUCTION

The modern world moves forward with an ever increasing demand for electricity. This growth in demand for electrical power is becoming a major challenge for electrical supply authorities worldwide with the present energy crises. Dwindling oil reserves, consequent rise in electricity costs and the mounting need to address concerns regarding climate change due to the emission of greenhouse gases (GHG) are forcing utilities to look at new concepts in electricity generation, transmission, distribution and utilization [1].

The present grid systems which have mostly been built over the last century are rapidly aging. Legacy power distribution systems are facing a tough challenge in meeting the needs of a rapidly evolving electricity market. Time is ripe for a smarter grid, capable of meeting the present and future power delivery requirements [1] - [3]. The need to reduce GHG emissions have ushered in the need for greener electricity generation and incorporating renewable energy sources into the energy mix. These complications demand a paradigm shift from large scale fossil fuel based sources towards small scale lower emission systems. The incorporation of smaller distributed generation (DG) units with maximum renewable energy penetration becomes vital in meeting these demands.

Centralized control systems are more powerful in controlling available resources in traditional power systems. However the current power network, with more renewable power sources with intermittent nature, dynamic behavior has increased considerably. In this context literature has recognized multi agent based systems gives more flexibility to the power network [4].

Grids with such DG systems require the ability to handle active generation and loads, which existing legacy system designs are unable to address [2].

Smaller "micro" grids comprising local generation is a concept put forward to cater this requirement [2]. These microgrids can be operated either in island mode, where the local loads are fully supplied by the local generation, or in grid connected mode, where the microgrid is either exporting or importing power from the main grid. With the local generation and loads coming into the picture, local distributed control is essential for the grid stability.

Therefore, these power systems have to become smarter with local distributed control capabilities and more reliable and robust in taking on renewable energy sources without losing stability and efficiency. Smarter microgrids are an answer to this situation. As an innovative method in providing this flexible distributed control requirement, Multi Agent Systems (MAS) are stepping forward [3]. MAS based applications can be found in several areas in the power system, such as electricity market applications [5] [6], smart grid related areas [7] [8] and power control applications [9] . Under power system control, load restoration [10] and load shedding [11] and are the key areas for microgird applications.

Agents in the MAS can use artificial intelligent (AI) techniques to attain the proper control of the microgrid to keep the system voltage and the frequency within the acceptable level [12] [13]. MAS architecture play a vital role in these applications [14] [15]. This paper proposes a novel MAS architecture which provides scalable and robust distributed control architecture for microgrid applications.

The MAS model is designed using the JADE [16] [17] platform and is implemented on a simulated test-bed in MATLAB/SIMULINK.

Rest of the paper is organized as follows. A brief introduction to microgrids and MAS is given in Section 2. A novel dual layered MAS architecture is proposed in Section 3. Simulation results of a conceptual microgrid operation based on the proposed architecture is discussed in Section 4. The conclusions are presented in Section 5.

## 2. MICROGRIDS AND MULTI AGENT SYSTEMS

Centralized, bulk generation facilities are giving way to more distributed and small scale generation systems. Distributed generation (DG) introduces the capability of embedding a wide range of low emission and potentially lower cost generation options. Such options can include internal combustion (IC) engines, micro-turbines and non-conventional renewable energy sources such as solar photo voltaic, wind, biomass, mini/micro-hydro and even waste-to-energy systems. These smaller generation technologies allow the sources to be placed optimally with respect to the loads, thus reducing emissions and transmission losses, while also allowing for local control instead of centralized dispatching [2].

Such DG systems provide the potential to take a sub-system approach in working with microgrids. During a disturbance in the main utility grid, local generation and their loads can be separated from the utility isolating the local loads from the disturbance, while maintaining the integrity of the main utility grid. This isolation is carried out at a single point of connection to the main utility known as point of common coupling. This capability, known as islanding, requires high reliability and flexibility from the microgrid. These requirements can be met by having peer-to-peer control and plug-and-play capabilities in each entity within the microgrid.

A conceptual microgrid model is shown in Figure 1. This can be either a small scale distribution level network or an electrical system of a building. Certain loads, which are critical, have local power sources. Non-critical loads have no local supply.

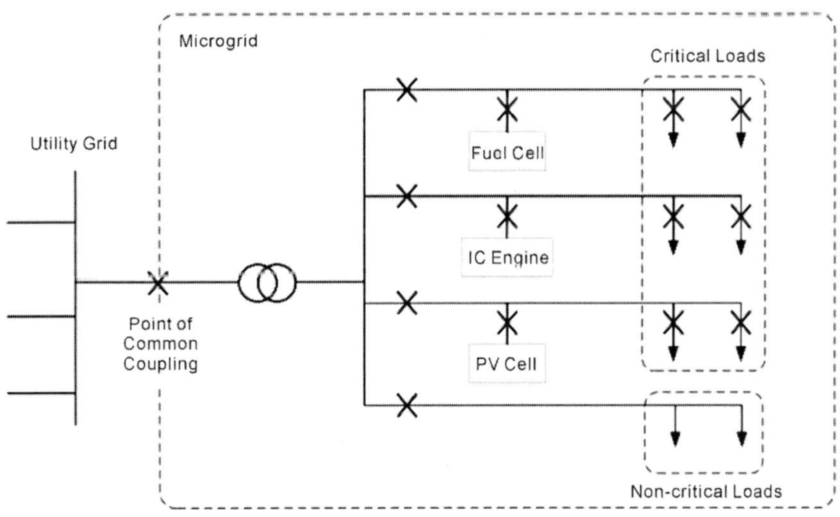


**Figure 1**. Conceptual microgrid model.

Once an upstream disturbance takes place on the main grid, the microgrid can be separated from the main grid at the point of common coupling and isolate the local loads from any outage, providing a higher level of service, with minimum effect to the stability and the integrity of the main grid. This capability of intentional islanding delivers a higher reliability than legacy power systems [2] . Thereafter, the microgrid has to manage the local loads to make sure that the most critical loads are supplied by the available limited local capacity. This requires priority based shedding of non-critical loads.

## 2.1. Multi Agent Systems (MAS)

With the need of peer-to-peer control, plug-and-play capabilities required in successfully implementing microgrids; more robust distributed control architectures are being required. Multi Agent Systems (MAS) have been successfully used to fulfill this requirement [18] - [21]. MAS are a collection of intelligent agents with limited local knowledge, but with enhanced communication capabilities, that can cooperate to achieve a global objective [22].

The notion of agent or agency had been developed for use mainly in computer science applications. A comprehensive comparison of the numerous definitions alongside with their benefits and drawbacks has been presented in [23]. The authors prefer the definition proposed by Wooldridge [22], as found in many of the recent literature. In [22], an agent is defined as a software entity inside an environment able to autonomously react to any changes. Here the notion of an agent, an environment and the ability of autonomy is defined. The environment suggests everything external to the agent. In order for an agent to behave in an environment it should, at least in part, be observable or alterable by the agent. In the microgrid paradigm, the environment and the agent are situated in the power system, which is observable through power-electronic sensors, and alterable through power-electronic devices i.e. closing or opening a breaker to reconfigure the power flow.

Definition in [22] differentiates agents from existing systems and objects in programming with the concept of flexible autonomy. Characteristics of agents are: Reactivity—ability to react to sensible changes in the environment; Pro-activity—make intuitive decisions and display goal-oriented behavior; and Social-ability—be able to communicate with other agents using a common language and initiate conversations.

Some existing control systems and software objects exhibit reactivity, but an agent with flexible autonomy differs from them by displaying a degree of intuition, alongside social ability [22]. Integration of two or more such intelligent agents coordinating to achieve a global objective by trying to fulfil local goals is a MAS.

Readers interested in a detailed guide to the concepts, methodologies, technical problems and the applicability of MAS in power systems applications are referred to [18]. An in depth discussion on the standards, tools, ontologies, design requirements and specifications can be found in [19].

## 2.2. MAS and Distributed Control of Microgrids

Agent systems are inherently distributable owing to their separation from their environment. An agent can be copied to a different location without losing reasoning capabilities or the goals. This allows for an agent to react in different ways to two different environments, given both environments support the intended tasks.

A basic issue for distributed generation is the technical difficulties related to control of a significant number of micro-sources. Research into implementing MAS for microgrid applications has been rising in the recent years and has mostly recently centered on distributed control, electricity trading, and power system restoration [20].

Intelligent agents can be assigned to each entity in a microgrid; i.e. circuit breaker, electricity generation unit, loads, users etc. This assignment will determine the MAS architecture. Several different architectures for microgrid applications have been proposed up to now i.e. [23] -[25] . Single layered architectures are the most pre-dominant type in the literature. For example, single layered architectures have been proposed for microgrid applications such as resolving real and reactive power mismatch [25] and load restoration [26]. In these single layered agent topologies, agents are tasked with multiple objectives. For instance, main controlling agent in these architectures is commonly tasked with islanding; load shedding, load management and process control simultaneously. This becomes a shortcoming in improving the scalability and in building robust distributed control models. Therefore, the appropriate delegation of objectives to secondary level agents improves the robustness of the model. Such secondary agents can be tasked separately with islanding control and load shedding, while overseen by a primary agent on the primary layer.

Several multiple level control architectures have been presented in the literature [27] - [29]. However, in these multi layered topologies, the layers address different strata of the distribution grid and not of the control architecture.

# 3. PROPOSED DUAL-LAYERED MAS

Figure 2 shows the conceptual design of the proposed dual layered control architecture for microgrids [30]. The proposed MAS system is developed using the JADE platform [16] [17] [31]. The model presents a distributed control architecture operating in two layers. The primary (strategic) layer comprises the Control agent who makes the run-time decisions. The control agent is in control of a secondary layer, overseeing the load control and microgrid connection control capabilities. The control agent is the center of the primary layer holding influence over the DG agent, controlling the LV and Load agents and interfacing with the User agent.

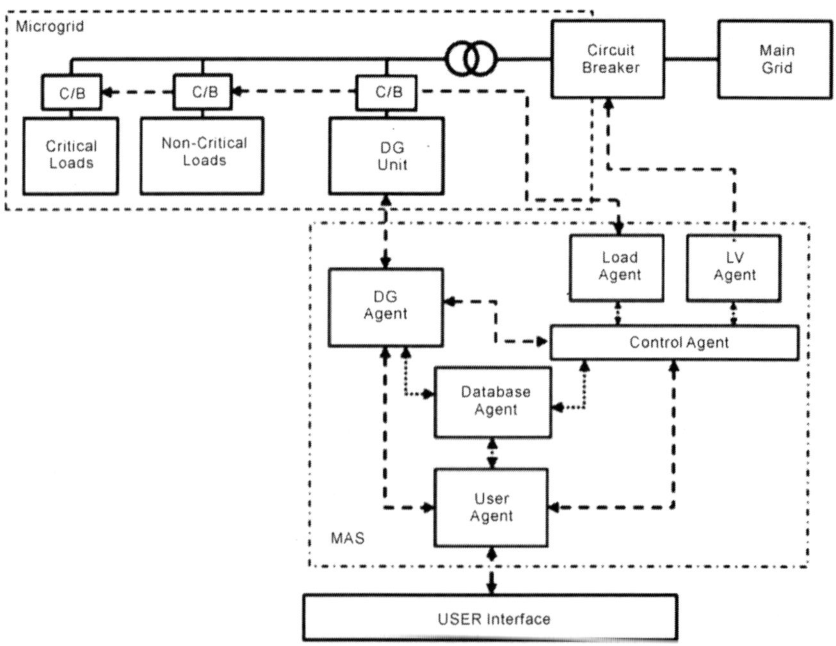


**Figure 2**. Dual layered control architecture.

The command hierarchy of the proposed dual layered MAS architecture is depicted in Figure 3. The secondary (tactical) layer contains the DG Agent, LV Agent, Load Agent and User Agent. This layer carries out the execution of the run-time decisions of the strategic level, which mainly comprises input/output operations. The DG agent collects several information related to the DG; 1) availability, 2) connection status, 3) power rating, 4) energy source availability and 5) cost of energy.

During an upstream fault, the LV agent islands the microgrid, by tripping the circuit breaker at the point of common coupling, if the Control agent allows it. During island operation the load shedding is carried out by the Load agents to ensure that power is delivered to the most critical loads on the priority basis set by the Control agent. This is implemented by Load agents opening circuit breakers at each of the controlled local loads.

The User agent behaves as the gateway for the user to interact with the system, to obtain real-time information and to set system goals. A utility agent acting in the form of a database agent is utilized to store system information and data and messages shared between the agents. The priority list used for load management is also hosted by the database agent. The database agent acts as the access point for data for all the agents and users.

The MAS is developed for the load management during the islanded phase. As local generation capacities will most definitely be limited with respect to local loads, load shedding becomes unavoidable during islanding. Proper load management during load shedding provides better protection to the most critical loads. By assigning priority values to the local loads, the most critical loads can be identified for protection during islanded mode. By allowing priority level updates during operation the control over load shedding/management can be improved.

Both these layers are ultimately supported by a service (utility) provided by a DB (Database) agent, running in the background. This utility agent stores system information and data and messages shared between the agents and acts as a data access point for all agents.

The agents will behave as hybrid agents [32] , capable of both reactive and deliberative actions. Reactive actions have the ability to respond quickly to emergencies of environment, while deliberative actions utilize the "belief, desire and intention" capability of agents [33] which allows them to adopt behaviors to achieve its objectives.

The control flow chart given in Figure 4 describes this load management during islanded mode. The user agent is allowed to update priority levels during runtime. By revising priority levels a user can restore power to a load that was previously considered non-critical.

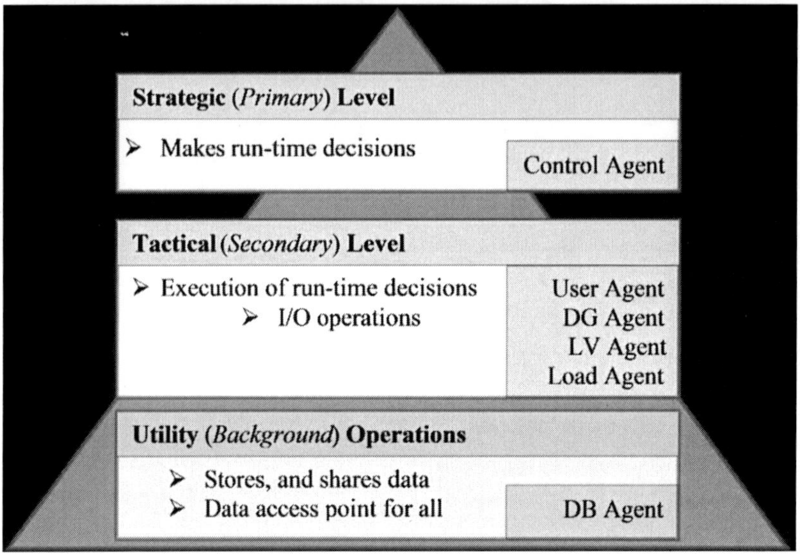


**Figure 3**. Command hierarchy of the proposed dual layered MAS architecture.

When the local capacity is deemed unable to meet the local demand, load shedding is initialized based on the priority values of the local loads. Then the load agents with lower priorities are advised to disconnect while maintaining supply to the most critical loads. When the priority list is updated by the user agent, it is hosted by the database agent and is broadcast to all the other connected agents. This triggers a re-evaluation of capacity and demand, and thereafter any necessary load shedding is carried out based on the revised/updated priority list.

The MAS controller allows the dynamic runtime update of load priorities for load management during a main grid outage. This allows a local grid operator to increase the priority of a currently disconnected local load, and reconnect it to the local grid depending on locally available capacity. A user can change the priority of load using the USER agent via a priority update message (Figure 3). This is updated in the database agent, who broadcasts a load priority update to the load agents.

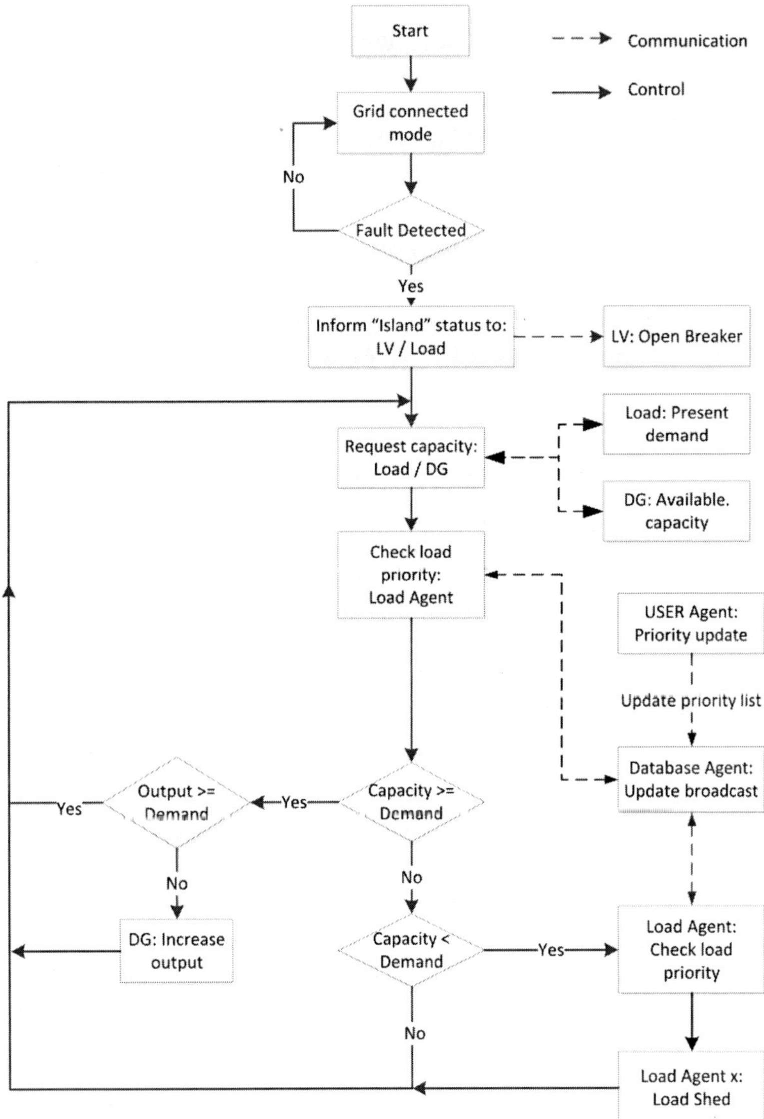


**Figure 4**. Control flow chart for load management.

If another "new" local load, needs to be added during islanded operation the same routine will be followed. The connection of the new load would depend on its priority level. If its priority precedes that of the current local loads, the new load will be connected to the load grid while the lower priority loads will be shed depending local capacity. If it has a lower priority, it will not be connected unless a user raises its priority via the USER agent.

The Proposed architecture, allows for self-organizing control. As each agent acts independently, once an agent is created or initiated it can seamlessly connect and logically interact with existing agents. This is done via the directory service provided by a utility agent acting as a database. A newly initiated agent can register in the directory and advertise its capabilities and requirements to the others. This provides self-organization capabilities to the control framework. Therefore, if one agent goes offline, the other agents can reorganize to cope with the loss. This ability increases the scalability and robustness of the controller.

# 4. SIMULATIONS AND RESULTS

Three simulations are carried out as case studies to verify the capability of the proposed MAS to provide control and protection for an inverter based microgrid. Within conventional power systems, frequency control is handled by rotating inertial masses of large generators. This causes a problem within smaller microgrids. Therefore, an inverter based system is selected, where the frequency control is handled by the inverter interface.

The multi agent system runs on a JAVA platform while the simulation is run on a simulating environment. Therefore, in order to establish communication between two domains, for sensing and control TCP/IP is used. A third party TCP/IP server [34] is utilized as an interface to MATLAB and it connects to another middle server which allows multiple simultaneous TCP/IP connections. Each agent requires a separate connection to the test-bed. The interface allows only a single connection at a time; therefore a separate middle server is required. This process is depicted inFigure 5.

The SIMULINK model for microgrid test bed designed for the simulations is shown in Figure 6. It consists a 480V, 15 kW embedded generator, acting as a DG unit, supplying part of the local demand, and two bulk local loads of 15 kW each. The loads are considered only at lumped load level, and connection/disconnection of part loads are not considered in this study. It is assumed that the DG units can operate at full capacity without fuel limitations during any outage.

In the inverter based microgrid is considered a grid interface unit comprising an inverter, low pass filter, Pulse Width Modulation (PWM), transformer and circuit breaker is used to connect a local DC source. The microgrid is connected to a main utility grid (11 kV) across a transformer (11 kV/400 V) and a main circuit breaker.

## 4.1. Intentional Islanding during Upstream Fault

The first case study is carried out to verify the effectiveness of the MAS in rapidly islanding the microgrid upon the detection of an upstream outage on the main grid. The test-bed used for the simulation is shown in Figure 6. It comprises a 15 kW DG unit producing 480 V, connected to a 400 V, 50 Hz microgrid, a transformer (480 V/400 V) connecting the DG to the microgrid, two 7.5 kW local loads, with a connection to a 11 kV main utility grid fed through another transformer (11 kV/400 V), all of which are connected across circuit breakers. Total local demand is 15 kW, 50% of which (7.5 kW) is supplied by the DG unit, while the rest is supplied by the main grid. It is assumed that the DG unit can instantly increase output using battery storage. An upstream fault is introduced to the system at t = 0.05 s into the simulation. At the beginning of the simulation breakers A, B, C and D are all closed. The test-bed shown in Figure 5 is simulated for 0.1 s. The simulation results depicting the successful islanding process are presented in Figure 7.

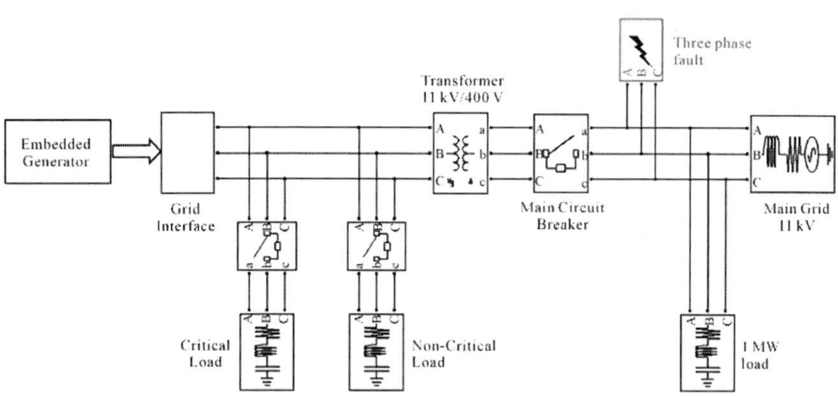


**Figure 5**. SIMULINK model for simulation test bed.

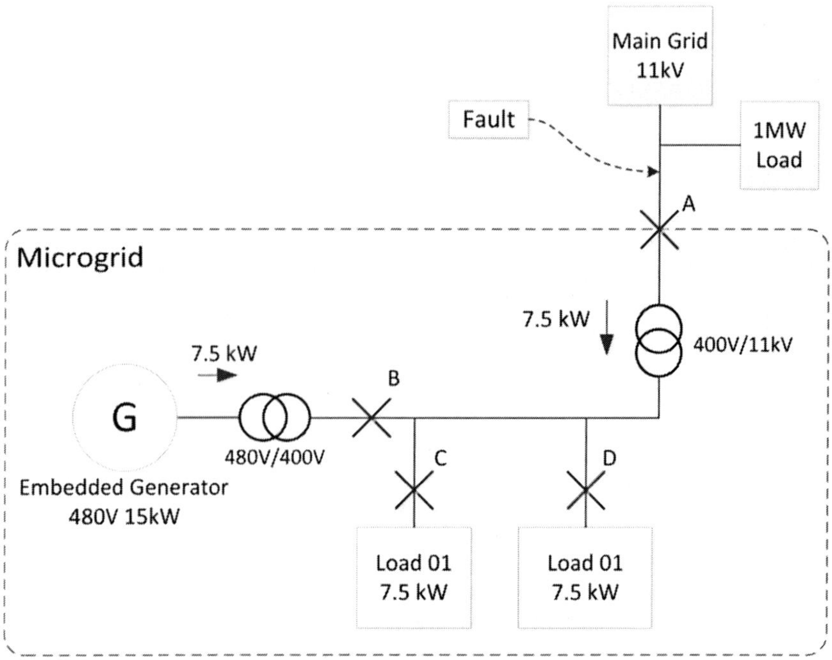


**Figure 6**. Simulation test-bed for case-study 1.

a) Upon detection of upstream fault:

When the upstream outage is detected at t = 0.05 s the control agent informs the LV agent and the Load agents to switch to the islanded mode operation. Upon receiving the islanding order the LV agent trips the main circuit breaker A, at the PCC, isolating the microgrid from the utility [Figure 7(a)]. The Control agent then queries the DG agent and the load agents regarding their available capacities that the DG can provide to the microgrid and the power requirements of the connected loads.

As the available maximum capacity of 15 kW is able to meet the total demand of 15 kW, the control agent commands the DG agent to increase DG output to match the demand. Soon as the microgrid is put to island mode the supply to the loads are maintained [see Figure 7(b)].

b) In islanded mode:

At t = 0.05s the microgrid is separated from the main grid and the local loads are secured. After the microgrid switches to island mode the total local demand is met by the DG unit by supplying 15 kW.

All agent operations are carried out rapidly, from detecting fault, opening the main breaker, connecting the local source and shedding loads, to stabilize the microgrid within 0.02 s. Therefore, the system is able to disconnect from the main utility grid and maintain the supply to the critical loads without suffering a brownout and/or blackout.

## 4.2. Securing Power to Critical Loads during Intentional Islanding

The second case study addresses the capability of the MAS to secure power to the critical loads during intentional islanding. This needs to happen immediately following the transition to islanded mode. If the available supply of the local DG unit is limited and insufficient to supply the total local demand, load shedding has to be done rapidly in order to secure supply to the most critical loads. The test-bed used for the simulation is shown in Figure 8. The test-bed is same as the one used in case study 1 (Figure 6), with only the following changes. Two local loads are 15 kW each with assigned priorities (p) of p = 0.7 and p = 0.4 respectively. The total local demand is 30 kW; initially the DG unit supplies 10 kW while the other 20 kW is provided from the main grid. Similar to case study 1, an upstream fault is introduced to the system at t = 0.05 s into the simulation. At the beginning of the simulation breakers A, B, C and D are all closed.

The test-bed shown in Figure 8 is simulated for 0.12 s. the simulation results shown in Figure 9depict the successful islanding and subsequent securing of power for the critical loads.

a) During grid connected mode:

While the microgrid is in grid-connected mode, the total demand is 30 kW, with 15 kW critical loads and 15 kW non-critical loads. During the grid-connected mode the DG unit provides 10 kW output and the other 20 kW is provided by the main grid. The main grid experiences a fault at t = 0.5 s.

b) Upon detection of upstream fault:

When the upstream outage is detected at t = 0.05 s the control agent informs the LV agent and the Load agents to switch to the islanded mode operation. Upon receiving the islanding order the LV agent trips the main circuit breaker A, at the PCC, isolating the microgrid from the utility [Figure 9(a)]. The Control agent then queries the DG agent and the load agents regarding their available capacities the DG can provide to the microgrid and the power requirements of the connected loads.

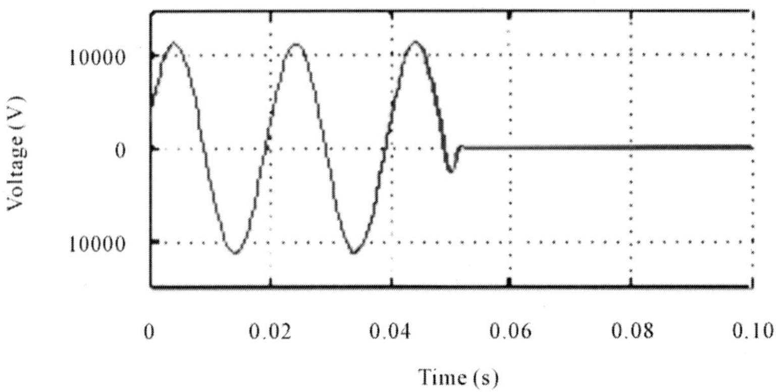


(a) Main grid supply voltage

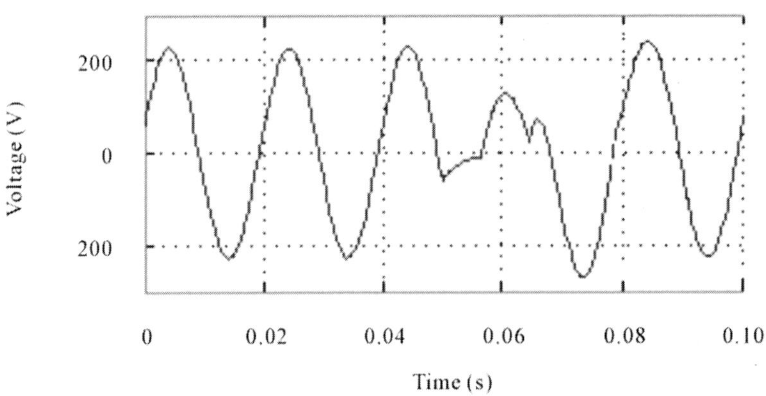


(b) Load voltage

**Figure 7**. Voltages measured during simulations (line-to-line) at (a) Line to line voltage across the non-critical load at "A"; (b) Line to line voltage at breaker "B".

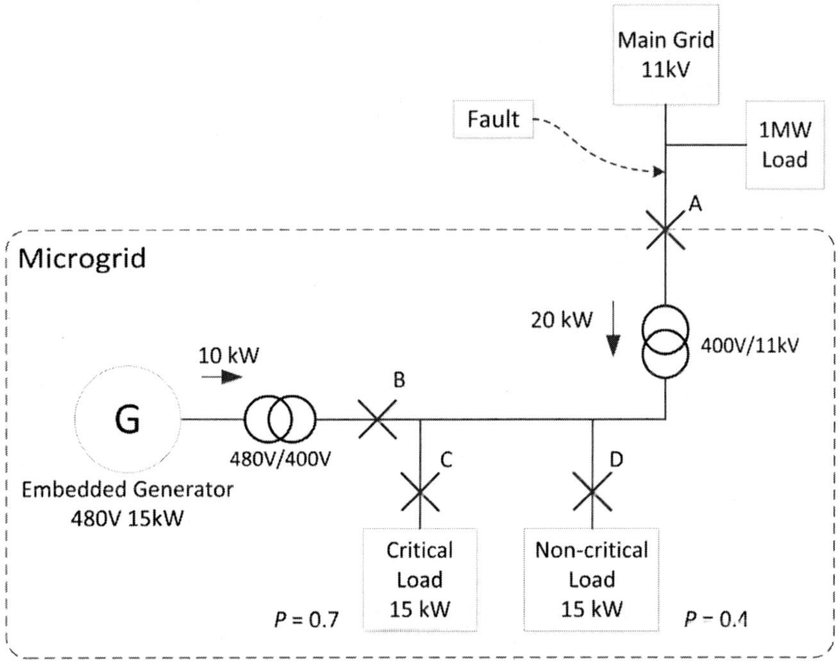


**Figure 8**. Simulation test-bed for case-study 2.

As the total load of 30 kW exceeds the available maximum capacity of 15 kW, the control agent commands the load agents to shed the 15 kW non-critical loads to match the DG capacity and the DG unit to increase output to 15 kW. Soon as the microgrid is put to island mode the supply to the critical loads is maintained [see Figure 9(b)]. The load agent at the non-critical loads sheds them from the system by opening breaker D [see Figure 9(c)].

c) In islanded mode:

At t = 0.05s the microgrid is separated from the main grid and the load agents balance the local demand. After the microgrid switches to island mode the total local demand is met by the embedded generator supplying 15 kW.

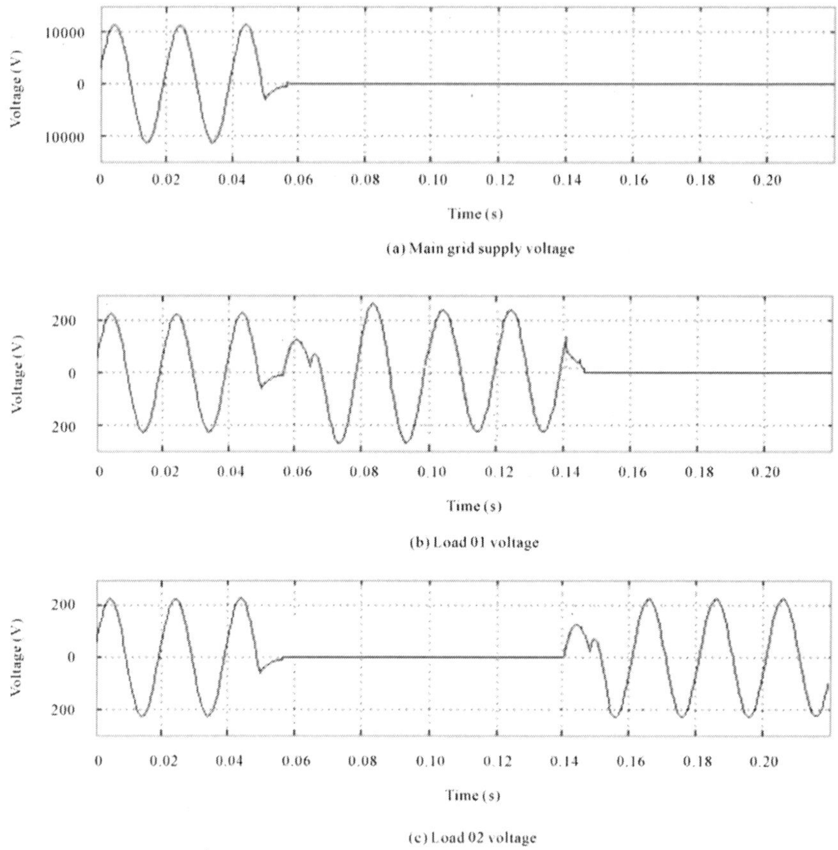


**Figure 9**. Voltages measured during simulations (line-to-line) at (a) Main grid voltage at breaker "A"; (b) Load 01 measured at "C": switches from high priority to low priority; (c) Load 02 measured at "D": switches from low priority to high priority.

## 4.3. Load Management during Islanded Mode Operation

The third case study addresses the capability of the MAS to provide load management capabilities during islanded operation. As the available local capacity is limited and insufficient to cater total local demand during islanded mode, non-critical (lower priority) loads will have to be dynamically shed in order to protect the stability of the microgrid. This simulation shows the ability of the MAS to allow for dynamic update of load priority levels. The simulation uses the test-bed shown in Figure 8.

The local loads are given the two same initial priority levels as shown in Figure 8. Local load 01 is given a priority level of p = 0.7, and load 02 is given a priority level p = 0.4. For this initial setting, their respective load agents designate load 01 as a higher priority load and load 02 as a lower priority load. At the beginning of the simulation, breakers A, B, C and D are all closed. An upstream fault is introduced to the system at t = 0.05 s into the simulation, and the microgrid switches to islanded operation. During islanded mode operation, the load 02 priority is raised to 0.9 at t = 0.14 s.

The test-bed shown in Figure 8 is simulated for 0.2 s. the simulation results shown in Figure 9depict the successful islanding and subsequent load management.

a) During grid connected mode:

While the microgrid is in grid-connected mode, the total demand is 30 kW, comprising two 15 kW local loads. During the grid-connected mode the embedded generator only provides 10 kW, while the other 20 kW is supplied by the main grid to the microgrid.

b) Upon detection of upstream fault:

Similar to case-study 1 and 2, the proposed MAS is able to successfully island the microgrid during the upstream outage within 0.02 s of detecting the fault in the main utility grid [see Figure 9(a)]. As the local generation capacity is unable to meet the total local demand, the MAS initiates load shedding upon initially set priority levels of the loads as in case study 2. This enables the MAS to maintain supply to the most critical load, load 01, by shedding the least critical load, load 02.

c) In islanded mode:

During the island operation, the user agent can revise the initial priority assignment. This can be pre-initiated or a user can change the priorities during the islanded mode.

At t = 0.14, the user agent revises the priority value of load 02 from p = 0.4 to p = 0.9 and communicates the "revise_request" to the load agent at load 02. This update is forwarded to other agents via the directory agent, and the control agent reinitiates load management procedures. As the DG unit is still unable to meet the total demand of 30 kW, and has to shed non-critical (lower priority) loads to maintain power to the new highest priority load, load 02.

Therefore, the control agent commands the load agents to shed the load 01 of 15 kW to match the DG capacity. Thus, the supply to the new most critical load, load 01 is provided by re-closing "D" [see Figure 9(c)]. The load agent at the current lower priority load 01 sheds them from the system by opening breaker "C" [see Figure 9(b)].

When the load priorities are revised by a user via the user agent, the MAS is able to reconfigure the system, in order to provide power to the current most critical load/s. The reconfiguration is also done within 0.02 s of the revision and successfully reconnects the new critical load. The results show the capability of the MAS to safely island and maintain the supply to its critical loads, while allowing for the critical loads to be dynamically revised.

All agent operations are carried out rapidly, from detecting fault, opening the main breaker, connecting the local source and shedding loads, to stabilize the microgrid within 0.02 s. the load reconnection after the priority update is done again within 0.02 s of the revision being sent from the user agent. The rapid transitions (within one cycle) ensure a high power quality with minimum effect to the loads. Therefore, the system is able to disconnect from the main utility grid and maintain the supply to the critical loads without suffering a brownout and/or blackout.

# 5. CONCLUSIONS

The proposed dual layered MAS is able to successfully island the microgrid during an upstream outage within 0.02 s of detecting the fault. As the local generation capacity is unable to meet the total local demand, the MAS initiates load shedding upon pre-set priority levels of the loads. This enables the MAS to maintain the supply to the most critical load/s by shedding least critical loads until DG capacity can meet the demand. The proposed MAS controller allows for dynamic/real time load management during islanded mode of operation. When the load priorities are revised by a user via the user agent, the MAS is able to reconfigure the system, in order to provide power to the new most critical load. The reconfiguration is also done within 0.02 s of the revision and successfully reconnects the new critical load. The rapid transitions (0.02 s) ensure a high power quality with minimum effect to the loads. The results show the capability of the MAS to safely island and maintain the supply to its critical loads, even if the critical loads are revised. These results validate the effectiveness of the MAS in controlling DG units to protect and control a microgrid.

The agent behaviour can further be developed using fuzzy/neural/heuristic algorithms. The long-term extensibility of MAS can be evaluated by studying the effect of increasing number of agents on controller stability and service latency. Another important area of future study will be the developing the requirements of communication network redundancy and data network security to prevent cyber-attacks and intrusions from taking down the smartgrid.

## ACKNOWLEDGEMENTS

The authors gratefully acknowledge the support provided by the Senate Research Council, University of Moratuwa (SRC/LT/2012/05).

## REFERENCES

1. Li, F., Qiao, W., Sun, H., Wan, H., Wang, J., Xia, Y., Xu, Z. and Zhang, P. (2010) Smart Transmission Grid: Vision and Framework. IEEE Transactions on Smart Grid, 1, 168-177.

2. Lasseter, R.H. and Piagi, P. (2004) Microgrid: A Conceptual Solution. Proceedings of the Power Engineering Specialists Conference (PESC'04), Aachen, 4285-4290.

3. Moslehi, K. and Kumar, R. (2010) A Reliability Perspective of the Smart Grid. IEEE Transactions on Smart Grid, 1, 57-64.

4. Dou, C., Hao, D., Jin, B., Wang, W. and An, N. (2014) Multi-Agent-System-Based Decentralized Coordinated Control for Large Power Systems. International Journal of Electrical Power & Energy Systems, 58, 130-139.

5. Markus, P., Wolfgang, K., Maytal, S. and Collins, J. (2013) A Reinforcement Learning Approach to Autonomous Decision-Making in Smart Electricity Markets. Machine Learning, 92, 5-39.

6. Salman, K., Rasheed, A.R., Leen-Kiat, S. and Sohrab, A. (2014) Multiagent Study of Smart Grid Customers with Neighborhood

Electricity Trading. Electric Power Systems Research, 111, 123-132.

7. Kahrobaee, S., Rajabzadeh, R.A., Leen-Kiat, S. and Asgarpoor, S. (2013) A Multiagent Modeling and Investigation of Smart Homes with Power Generation, Storage, and Trading Features. IEEE Transactions on Smart Grid, 4, 659-668.

8. Wang, X., Zhang, P., Wang, Z., Dinavahi, V., Chang, G., Martinez, J.A., Davoudi, A., Mehrizi-Sani, A. and Abhyankar, S. (2013) Interfacing Issues in Multiagent Simulation for Smart Grid Applications. IEEE Transactions on Power Delivery, 28, 1918-1927.

9. Muller, S.C., Hager, U. and Rehtanz, C. (2014) A Multiagent System for Adaptive Power Flow Control in Electrical Transmission Systems. IEEE Transactions on Industrial Informatics, 10, 2290-2299.

10. Horbani, J., Chouhan, S., Choudhry, M.A. and Feliachi A. (2013) Hybrid Multi Agent Approach for Power Distribution System Restoration. Proceedings of the IEEE Energytech, Cleveland, 1-5.

11. Lim, Y., Kim, H.M. and Kinoshita, T. (2014) Distributed Load-Shedding System for Agent-Based Autonomous Microgrid Operations. Energies, 7, 385-401.

12. Sanjari, M.J. and Gharehpetian, G.B. (2014) Game Theoretic Approach to Cooperative Control of Distributed Energy Resources in Islanded Microgrid Considering Voltage and Frequency Stability. Neural Computing and Applications, 25, 343-351.

13. Sanjari, M.J. and Gharehpetian, G.B. (2014) Incomplete Information Based Decentralized Cooperative Control Strategy for Distributed Energy Resources of VSI-Based Microgrids. Neural Computing and Applications, 25, 189-197.

14. Niannian, C. and Mitra, J. (2014) A Multi-Level Control Architecture for Master-Slave Organized Microgrids with Power Electronic Interfaces. Electric Power Systems Research, 109, 8-19.

15. Ling, W. and Liu, D. (2014) A Distributed Fault Localization, Isolation and Supply Restoration Algorithm Based on Local Topology. International Transactions on Electrical Energy Systems, in Press.

16. JADE Agent Development Toolkit (2014) http://jade.tilab.com

17. Bellifemine, F.L., Caire, G., Poggi, A. and Rimassa, G. (2003) A White Paper EXP in Search of Innovation. http://jade.tilab.com/papers/2003/WhitePaperJADEEXP.pdf

18. Mcarthur, S.D.J., Davidson, E.M., Catterson, V.M., Dimeas, A.L., Hatziargyriou, N.D., Ponci, F. and Funabashi, T. (2007) Multi-Agent Systems for Power Engineering Application Part I: Concepts, Approaches, and Technical Challenges. IEEE Transaction on Power Systems, 22, 1743-1752.

19. Mcarthur, S.D.J, Davidson, E.M., Catterson, V.M., Dimeas, A.L., Hatziargyriou, N.D., Ponci, F. and Funabashi, T. (2007) Multi-Agent Systems for Power Engineering Applications—Part II: Technologies, Standards, and Tools for Building Multi-Agent Systems. IEEE Transaction on Power System, 22, 1753-1759.

20. Kulasekera, A.L., Gopura, R.A.R.C., Hemapala, K.T.M.U. and Perera, N. (2012) A Review on Multi-Agent Systems in Microgrid Applications. Proceeding of the Innovative Smart Grid Technologies, India, 173-177.

21. Kumar, N.H.S.V.S. and Doolla, S. (2013) Multiagent-Based Distributed-Energy-Resource Management for Intelligent Microgrids. IEEE Transactions on Industrial Electronic, 60, 1678-1687.

22. Wooldridge, M. and Weiss, G. (1999) Intelligent Agents, Multi-agent Systems. Cambridge, 3-51.

23. Franklin, S. and Graesser, A. (1996) Is It an Agent or Just a Program? Proceeding of the 3rd International Workshop Agent Theories, Architectures, and Languages, New York, 21-35.

24. Cai, N. and Mitra, J. (2010) A Decentralized Control Architecture for a Microgrid with Power Electronic Interfaces. Proceeding of the North American Power Symposium, USA, 1-8.

25. Foo Eddy, Y.S., Gooi, H.B. and Chen, S.X. (2015) Multi-Agent System for Distributed Management of Microgrids. IEEE Transactions on Power System, 30, 24-34.

26. Xu, Y. and Liu, W. (2011) Novel Multiagent Based Load Restoration Algorithm for Microgrids. IEEE Transactions on Smart Grid, 2, 152-161.

27. Zhang, J., Ai, Q., Jiang, C.W., Wang, X.G., Zheng, Z.H. and Gu, C.H. (2009) The Application of Multi Agent System in Microgrid Coordination Control. Proceeding of the International Conference on Sustainable Power Generation and Supply, Nanjing, 1-6.

28. Zheng, W. and Cai, J. (2010) A Multi-Agent System for Distributed Energy Resources Control in Microgrid. Proceeding of the 5th International Conference on Critical Infrastructure (CRIS), Beijing, 1-5.

29. Aquino-Lugo, A.A., Klump, R. and Overbye, T.J. (2011) A Control Framework for the Smart Grid for Voltage Support Using Agent-Based Technologies. IEEE Transactions on Smart Grid, 2, 173-180.

30. Kulasekera, A.L., Gopura, R.A.R.C., Hemapala, K.T.M.U., Perera, N. and Pallegedara, A. (2012) Dual Layered Multi Agent System for Intentional Islanding Operation of Microgrids. Proceeding of the 17th International Symposium on Artificial Life and Robotics, Oita, 994-997.

31. Bellifemine, F., Caire, G. and Greenwood, D. (2007) Developing Multi-Agent Systems with JADE. John Wiley & Sons Ltd., 51-111.

32. Dou, X.C. and Liu, B. (2013) Multi-Agent Based Hierarchical Hybrid Control for Smart Microgrid. IEEE Transactions on Smart Grid, 4, 771-778.

33. Georgeff, M., Pell, B., Pollack, M., Tambe, M., Wooldridge, M., Müller, J.P., Rao, A.S. and Singh, M.P. (1999) The Belief-Desire-Intention Model of Agency. Proceeding of the Intelligent Agents, 1-10.

34. Zimmermann, W. (2011) IOlib-Hardware Input/Output Library Module for Matlab and Simulink. http://www.it.hs-esslingen.de /~zimmerma /software/ index_uk.html

# 9

# CHAPTER

# A MULTI-AGENT FRAMEWORK FOR PACKET ROUTING IN WIRELESS SENSOR NETWORKS

**Dayon Ye [1,*], Minji Zhang [2] and Yu Yang [1]**

*[1]School of Software and Electrical Engineering, Swinburne University of Technology, Melbourne 3122, Australia*

*[2]School of Computer Science and Software Engineering, University of Wollongong, Wollongong 2522, Australia*

## ABSTRACT

Wireless sensor networks (WSNs) have been widely investigated in recent years. One of the fundamental issues in WSNs is packet routing, because in many application domains, packets have to be routed from source nodes to destination nodes as soon and as energy efficiently as possible. To address this issue, a large number of routing approaches have been proposed. Although every existing routing approach has advantages, they also have some disadvantages. In this paper, a multi-agent framework is proposed that can assist existing routing approaches to improve their routing performance. This framework enables each sensor node to build a cooperative neighbour set based on past routing experience. Such cooperative neighbours, in turn, can help the sensor to effectively relay packets in the future. This framework is independent of existing routing approaches and can be used to assist many existing routing approaches. Simulation results demonstrate the good performance of this framework in terms of four metrics: average delivery latency, successful delivery ratio, number of live nodes and total sensing coverage.

**Keywords:** multi-agent systems; wireless sensor networks; packet routing

# 1. INTRODUCTION

Due to recent technological advances, the manufacturing of small, low-power, low-cost and highly integrated sensors has become technically and economically feasible. These sensors are generally equipped with sensing, data processing and communicating components. Thus, such sensors can be used to measure ambient conditions in the environment surrounding them and then transform these measurements into signals. The signals can be processed further to reveal some properties about objects located in the vicinity of the sensors. The sensors then send such collected data, usually via a radio transmitter, to a command centre (also known as a sink or a base station) either directly or via several relaying sensors. A large number of these sensors can be networked in many applications that require unattended operations, hence producing a WSN. Currently, there are various applications of WSNs. These applications include battlefield surveillance [1], video capture, processing and communication [2,3], *etc*. Typically, WSNs contain hundreds or thousands of sensors, and these sensors have the ability to communicate with each other. Thus, packet routing is an important issue in WSNs. In addition, the routing issue in WSNs is more challenging than in other types of wireless networks, e.g., mobile *ad hoc* networks or cellular networks [4], because the power energy, storage capability and processing capacity of each sensor are usually constrained. Moreover, unlike general computer networks, where the destination of data is variable, in WSNs, the destination of nearly all data is a sink node. Therefore, designing routing approaches in such sensor networks needs careful resource management.

Over the last decade, many approaches have been proposed in order to solve the routing problem in WSNs [5,6]. These approaches have taken into account the inherent features of WSNs along with particular applications and architecture requirements. It is not a simple task to find and maintain routes in WSNs due to energy constraints and sensor failures. To mitigate energy consumption, existing routing approaches proposed for WSNs have employed some well-known routing techniques from other wireless networks and special techniques in WSNs, such as data aggregation, clustering, different node role assignment, *etc*. According to Al-Karaki and Kamal [7], almost all of the routing approaches can be classified into three major categories: flat routing approaches, hierarchical routing approaches and location-based (also known as geographic) routing approaches, although there are a few exceptional ones based on network flow or quality of service (QoS) awareness [8]. Routing approaches belonging to each of the three categories have advantages and disadvantages. In flat routing approaches, all sensors play the same role, and each sensor could be a source node or a relay node. Hence, the routing delay can be minimised, if the routing algorithm is carefully designed. The energy consumption is an outstanding problem, however, compared with the routing approaches belonging to other two categories, because sensors lack the knowledge of the global network, and thus, broadcasting is often required or data have to be repeatedly transferred in the network. Hierarchical routing approaches aim to cluster sensor nodes so that cluster heads can aggregate and reduce data to be transmitted to the sink node in order to save energy. Because packet transmission is the most energy expensive task [9], the reduction of packets is vital for the overall network. However, the packet routing delay will be lengthened, as cluster heads have to wait for a period for sensors in their clusters to transfer packets to them for aggregation. Location-based routing approaches use the position information to relay the packets to the desired regions rather than the whole network. Location-based routing approaches have good scalability, since the information of the whole network is not required. Such location-based approaches, however, often need a device, such as a GPS, to provide the necessary position information. Furthermore, the cost and weight of a sensor will be increased if each sensor is equipped with such a device. Although several researchers have suggested using a virtual coordinate to provide the position information [10], this would incur extra energy consumption due to the communication process, which is necessary for acquiring such position information. In addition, to obtain such virtual coordinate-based position information, global information has to be used, e.g., the maximum ID among all sensors with the maximum hop distance from a specific node.

As can be seen from the above, since all existing routing approaches have some disadvantages, it is more important to develop an assistant tool to enhance existing routing approaches than to develop a new routing approach. To the best of our knowledge, this paper is the first one to propose a multi-agent framework that can be used to assist and enhance many existing routing approaches. The merits of this framework include that: (i) it is independent of individual existing routing approaches, so it can assist many existing routing approaches and enhance their performance; (ii) it is decentralised and operated based only on local information, and it does not need to change the physical network topology; and (iii) it does not only suit static WSNs, but also works well in mobile and open WSNs, where sensors are mobile and new sensors may join or existing sensors may leave the network.

The rest of the paper is organised as follows. In the next section, a detailed literature review is provided, which includes routing approaches belonging to the aforementioned three categories, especially those approaches developed after the publication of the survey papers in [4,7]. Then, the proposed framework is described in detail in Section 3. A simulation regarding the performance of the proposed framework is given in Section 4. Finally, this paper is concluded in Section 5.

# 2. LITERATURE REVIEW

As the routing issue is one of the most significant issues in WSNs, there is a large number of published prominent studies that focus on this issue. Akkaya and Younis [4] and Al-Karaki and Kamal [7] provided a thorough survey on routing approaches nearly at the same time. Thus, in this section, our review focuses primarily on the routing approaches developed after their survey papers were published.

## 2.1. Flat Routing Approaches

The first category of routing approaches is multi-hop flat routing. In such routing approaches, each sensor node typically plays the same role, and sensor nodes collaborate to perform sensing tasks.

Flooding and gossiping [11] are two classical approaches to relay packets in sensor networks without the need for any routing algorithms or topology maintenance. In flooding, each sensor that receives a packet broadcasts the packet to all of its neighbours, and this process continues until the packet arrives at the destination or the maximum number of hops for the packet is reached. On the other hand, gossiping is a slightly enhanced version of flooding where the receiving node sends the packet to a randomly-selected neighbour, which picks another random neighbour to forward the packet to, and so on, until the destination or the maximum hop is reached.

Heinzelman *et al.* [12,13] proposed a family of adaptive protocols, called sensor protocols for information via negotiation (SPIN). The SPIN family is designed to address the deficiencies of classical routing approaches, such as flooding and gossiping, which waste much energy and bandwidth when sending extra and unnecessary copies of packets by sensors that cover overlapping areas. SPIN consists of three steps. First, when a sensor has a data packet to transfer, it broadcasts the metadata of the data packet to its neighbours. The size of the meta packets is much smaller than that of the data packet. Then, if any neighbours are interested in the data packet, they will send back a request message to the sensor. Once the sensor receives the request message, it will transfer the data packet to those neighbours that are interested in that packet.

Rogers *et al.* [14] devised an energy-aware self-organised routing algorithm for WSNs. Their algorithm enables individual sensors to follow locally selfish strategies, which, in turn, result in a framework of a routing network with desirable global properties. In addition, the algorithm can adaptively deal with the changing sensor numbers and network topology.

Beraldi *et al.* [15] studied the lowest latency path problem and then devised a forwarding protocol based on biased random walks, where sensors use only local information about their neighbours and their next active period to make forwarding decisions. Specifically, their protocol, lukewarm potato, is a compromise between the shortest path forwarding and hot potato forwarding. Hot potato forwarding is a greedy approach to the fast propagation of a message toward the sink: a node, upon receiving a message, forwards the message to the neighbour that is the first to become active.

Quang and Kim [16] devised a gradient routing algorithm that uses two-hop information for industrial WSNs to enhance real-time performance with energy efficiency. The routing algorithm is based on the number of hops to the sink instead of distance. Specifically, the routing algorithm applied two-hop velocity-based routing [17] to a gradient-based WSNs so as to reduce both energy consumption and end-to-end delay.

Niu *et al.* [18] introduced a routing enhancement scheme. Their scheme first finds a robust guide path during the route discovery phase. Then, along this guide path, data packets are greedily forwarded toward the destination through nodes' cooperation without utilising location information. The forwarding procedure is based on the priorities of forwarding candidates, where nodes with the highest priorities will be picked up to forward data packets.

## 2.2. Hierarchical Routing Approaches

Hierarchical or cluster-based routing approaches, originally proposed in wired networks, are well-known techniques with special advantages related to efficient communication. The concept of hierarchical routing is also utilised to perform energy-efficient routing in WSNs. In a hierarchical architecture, high-energy sensors can be used to process and transmit information, whereas low-energy sensors can be employed to perform sensing tasks in the proximity of the target. Hierarchical routing is usually a two-layer routing approach, where one layer is composed of cluster heads and the other layer consists of cluster members. Cluster members are responsible for sensing tasks, while cluster heads are used to aggregate packets and transmit the aggregated packets to the sink node. Since the size of aggregated packets is much less than that of raw packets, the energy consumption used for packet transmission can be significantly reduced, and thus, the overall network lifetime can be prolonged.

The most famous hierarchical routing protocol is LEACH (low-energy adaptive clustering hierarchy) [19]. LEACH randomly selects a few sensor nodes as cluster heads and rotates this role to evenly distribute the energy load among the sensors in the network. In LEACH, cluster heads compress packets arriving from cluster members that belong to the respective clusters. Cluster heads then send aggregate packets to a sink in order to reduce the number of packets that must be transmitted to the sink.

In [20], an enhancement over the LEACH protocol was proposed, *i.e.*, power-efficient gathering in sensor information systems (PEGASIS). PEGASIS is a near optimal chain-based protocol. The basic idea is that in order to extend network lifetime, sensor nodes need to communicate only with their closest neighbours, and these neighbours take turns in communicating with the sink. Unlike LEACH, PEGASIS avoids cluster formation and uses only one sensor node in a chain instead of multiple nodes to transmit the sink.

Tsai [21] presented a coverage-preserving routing protocol that was a modified version of LEACH. Unlike most routing protocols, the aim of his protocol is to preserve sensing coverage when some sensor nodes are no longer available, because, for example, their energy is used up. Tsai considered that simply increasing the lifetime of sensor nodes might not automatically achieve good sensing coverage. Thus, unlike LEACH, Tsai's protocol considered not only energy consumption, but also sensing coverage when clusters were formed.

Valentini *et al.* [22] devised a non-dominated algorithm to improve a simple hybrid routing protocol (SHRP) [23] in choosing the best route towards the sink node. Their algorithm allows simultaneous analysis of energy- and latency-related metrics to generate a Pareto-optimal solution, which has better time convergence and reliability than SHRP.

Mottola and Picco [24] developed an adaptive energy-aware routing protocol that was expressly designed for many-to-many communication, *i.e.*, simultaneously routing from multiple sources to multiple sinks. To increase network lifetime, their protocol minimises the number of nodes involved in many-to-many routing and balances forwarding load, reduces the amount of redundant information flowing in the network and decreases contention on the wireless medium and packet collisions.

## 2.3. Location-Based Routing Approaches

In this kind of routing, also known as geographic routing, sensor nodes are addressed in terms of their locations. The distance between neighbouring nodes can be estimated on the basis of incoming signal strength. The relative coordinates of neighbouring nodes can be obtained by exchanging information between neighbours. Alternatively, the location of sensor nodes may be achieved directly by communicating with a satellite using GPS if each node is equipped with a small low-power GPS receiver.

GAF (geographic adaptive fidelity) [25] is an energy-aware location-based routing algorithm designed primarily for mobile *ad hoc* networks, but may be applicable to WSNs, as well. In GAF, the network area is first divided into fixed zones. These zones then form a virtual grid. Inside each zone, nodes collaborate with each other to play different roles, and each node uses its GPS-indicated location to associate itself with a point on the virtual grid.

Yu *et al.*'s protocol [26], *i.e.*, GEAR (geographic and energy-aware routing), uses energy-aware and geographically-informed neighbour selection heuristics to route a packet toward the destination region. The key idea is to restrict the number of messages in directed diffusion by considering only a certain region rather than sending the messages to the whole network. Thus, the energy for transmission can be conserved.

Huang *et al.* [27] presented an energy-aware and interference-sensitive geographic routing protocol that focuses on minimising the total network energy consumption and reducing interference. Their routing protocol adaptively uses an anchor list to guide packet delivery and selects the minimum-interference link from the energy-optimal relay region for packet delivery.

Awad *et al.* [28] proposed the virtual cord protocol (VCP), which exploits virtual coordinates to provide efficient and failure-tolerant routing and packet management in sensor networks. Specifically, VCP uses two mechanisms to find paths to nodes and associated packet items. First, it relies on the virtual cord that always provides a path toward the destination. Second, locally available neighbourhood information is exploited for greedy routing.

Petrioli *et al.* [29] presented a geographic routing protocol that considered three important, but often neglected design challenges: routing around connectivity holes, resilience to localisation errors and efficient relay selection. In addition, their protocol integrates awake/asleep schedules, MAC, routing, traffic, load balancing and back-to-back transmissions. During the forwarding procedure, a sender checks the availability of its awake neighbouring nodes and channels. Then, the sender chooses the best relaying node based on its neighbouring nodes' forwarding performance. If several nodes have the same performance, positive geographic advancement toward the sink is used to discriminate among them. Such joint optimisation can make the protocol robust and realistic.

After reviewing related work, it can be seen that each approach has its own strengths and limitations. We propose, therefore, that it is important to devise a framework that can be applied to many existing routing approaches in order to improve their performance. This paper proposes such a framework, which will be described in detail in the next section.

## 3. THE MULTI-AGENT FRAMEWORK

In order to realise the framework, a two-layer architecture is introduced, where the first layer is the wireless sensor network and the second layer is a multi-agent cooperation network, as shown in Figure 1.

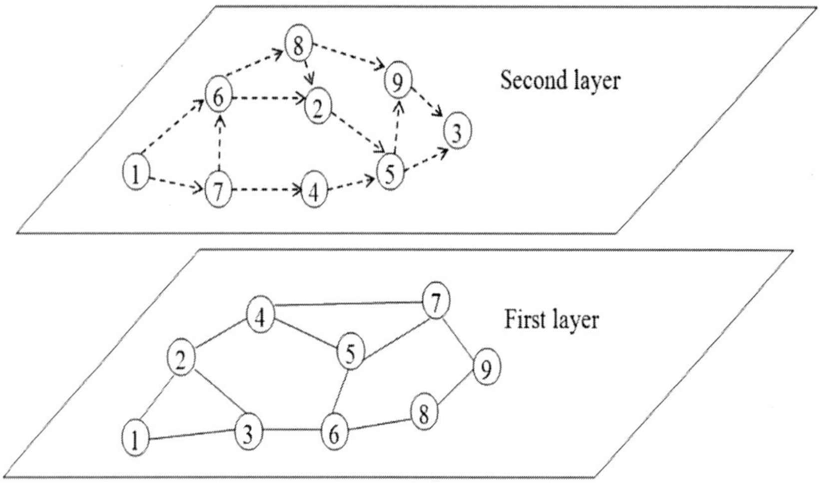


**Figure 1.** A sample two-layer architecture.

In the first layer, the wireless sensor network layer, sensors are connected by a wireless medium, such as infrared devices or radio, which is represented by the solid lines connecting the two sensor nodes. The first layer is the physical network, which really exists. In the second layer, each sensor is modelled as an agent, and agents are linked by a cooperative relation, represented as dashed lines. In the rest of the paper, the two terms, sensor and agent, are used interchangeably. The second layer, the multi-agent cooperation network layer, is an abstract network, which does not really exist. The multi-agent cooperation network is formed on the basis of the sensors' past cooperation. For example, in Figure 1, if many packets sent from Sensor 1 have to be forwarded by Sensor 7, Sensor 1 may add Sensor 7 as one of its cooperative neighbours. Then, in the future, if Sensor 1 has packets to be sent, it first sends the packets to Sensor 7, and Sensor 7 then forwards the packets to the destinations for Sensor 1. For the process that Sensor

1 sends packets to Sensor 7 and Sensor 7 forwards packets to the destinations, Sensors 1 and 7 can use any existing routing approaches. It can be seen that the multi-agent cooperation network layer is used to guide the packet routing process in the wireless sensor network layer in order to improve the routing efficiency, instead of becoming a concrete routing protocol that directly operates in the wireless sensor network layer. Thus, in this case, many existing routing protocols can be employed in the wireless sensor network layer. Such a two-layer architecture is necessary, because in the first layer, the topology of a WSN cannot be altered (unless some nodes move or leave the network, or new nodes join the network), and hence, the proposed framework has to work in the second layer, the multi-agent cooperation network layer, which enables each agent to keep the most useful cooperative neighbours and, in turn, assists existing routing approaches to achieve better performance.

In this paper, the sensor network is defined as a graph, $G = (V,E)$, where $V = \{v_1,...,v_n\}$ is the set of $n$ sensors in the network and $E \subseteq V \times V$ is a set of physical links. Two sensors, $v_i$ and $v_j$, are physical neighbours if $(v_i, v_j) \in E$and they can directly transmit packets between each other. $N_i = \{v_j|(v_i, v_j) \in E\}$ indicates sensor $v_i$'s physical neighbour set. Similarly, the cooperation network is defined as a graph, as well, $G' = (A, E')$, where $A$ is the set of agents and $E' \subseteq V \times V$ is a set of cooperation links. Each agent represents a sensor node. Two agents, $ag_i$ and $ag_j$, are cooperative neighbours if $(ag_i, ag_j)$ $E'$. N'i={agj|(agi,agj)∈E'} indicates agent $ag_i$'s cooperative neighbour set. It should be noticed that physical neighbours are symmetrical, namely that if $(v_i, v_j) \in E$, then $(v_j,v_i) \in E$ holds, but cooperative neighbours are asymmetrical, namely that when $(ag_i, ag_j) \in E'$, it is not necessary that $(ag_j, ag_i) \in E'$ holds. Initially, each sensor's physical neighbour set is identical to its cooperative neighbour set (although later, its cooperative neighbour set may be adapted).

Before devising the framework, it is necessary to introduce some evaluation indices to estimate each sensor, as well as the sensor network. These indices include energy consumption, storage consumption and sensing coverage.

## 3.1. Energy Consumption

Energy consumption relies on how many packets are being transmitted and the transmission distance of each packet. For a single sensor, say $v_i$, its energy consumption for transmitting a $k$-bit message to a receiver, say $v_j$, with a distance $d$ away can be calculated using Equation (1) [19].

$$Ene_{T_i}(k,d) = k\big(E_{elec} + \epsilon_{amp} d^{\beta}\big) \quad (1)$$

where $E_{elec}$ represents the energy being dissipated to operate the transmitter or receiver circuitry per bit and $\epsilon_{amp}$ denotes the energy dissipation of the transmitter power amplifier for transmitting a bit to a receiver with a distance of $d = 1$ unit away, and $\beta$ is a constant coefficient that denotes the path loss exponent.

For the receiver, $v_j$, its energy consumption for receiving the $k$-bit message is:

$$Ene_{R_j}(k) = kE_{elec} \quad (2)$$

Thus, for the whole sensor network, its energy consumption in a pre-defined period is the sum of the energy consumption caused by sensor nodes transmitting and receiving data packets in that period.

Certainly, other factors, such as idle listening, overhearing and synchronisation, can also incur energy consumption. To simplify the model, such factors are not taken into consideration.

## 3.2. Storage Consumption

Storage consumption involves two types of consumption. The first type of storage consumption is for maintaining packets in sensors' waiting lists. For example, if a sensor has several packets to send, but in a time slot, it can only send one packet, it has to keep the remaining packets in its waiting list and transmit them in the succeeding time slots. It is important for a sensor to keep as few packets in its waiting list and as short of a period of time as possible, so that both the sensor's storage consumption and the packet transmission latency can be reduced. For a single sensor, $v_i$, its storage consumption for maintaining packets can be calculated using Equation (3).

$$Stor_i^{(1)} = S^{(1)} \cdot \Big(\sum_{j=1}^{|W_i|} t_{ij}^{(1)}\Big) \quad (3)$$

where $W_i$ denotes the waiting list of sensor $v_i$, $t_{ij}^{(1)}$ demonstrates the number of time slots during which sensor $v_i$maintains a packet $j$ and $S^{(1)}$ is a constant to represent the storage coefficient for keeping packets.

The second type of storage consumption for an individual sensor is to keep (both physical and cooperative) neighbours, which can be calculated by employing Equation (4).

$$Stor_i^{(2)} = S^{(2)} \cdot (\sum_{j=1}^{|N_i|} t_{ij}^{(2)} + \sum_{l=1}^{|N_i'|} t_{il}^{(3)}) \tag{4}$$

where $N_i$ and N'i are the physical and cooperative neighbours of $v_i$, respectively, which have been described above; t(2)ij indicates the time slots during which sensor $v_i$ keeps $v_j$ as its physical neighbour; t(3)il indicates the time slots during which sensor $v_i$, *i.e.*, agent $ag_i$, keeps $ag_l$ as its cooperative neighbour; $S^{(2)}$ is a constant to represent the storage coefficient for keeping neighbours.

he reason for taking time into account when computing storage consumption is that in Equation (3), the time used to keep packets can indirectly reflect transmission latency, and in Equation (4), the time used to keep neighbours can, to some extent, reflect the computational overhead associated with taking each neighbour into consideration whenever a sensor sends a packet.

Likewise, for the whole sensor network, the storage consumption is the sum of each sensor's storage consumption.

## 3.3. Sensing Coverage

Sensing coverage is also an important property of a sensor network, although this property was overlooked by many researchers. As described in [21], saving unnecessary power consumption can prolong the lifetime of sensor nodes, but does not necessarily imply that a better sensing coverage can be achieved. Hence, it is necessary to consider sensing coverage as a separate property.

In many applications, sensor nodes are deployed randomly over the entire desired area. In this way, the sensing areas of different nodes may partially overlap. When a local area has a much higher node density than the average node density, a target location in that area may be covered by multiple sensors. On the other hand, if the node density of a local area is much lower than the average node density, a target location is more likely covered by only one sensor. The normalised effective sensing area, $\eta$, of a sensor, $v_i$, is calculated using Equation (5) [21].

$$\eta = \eta_0 + \sum_{m=1}^{\infty} \frac{\eta_m}{m+1} \tag{5}$$

where $\eta_0$ is the percentage of $v_i$'s sensing area that is covered by $v_i$ only and $\eta_m$ represents the percentage of $v_i$'s sensing area that is covered by $v_i$ and any other $m$ neighbouring nodes. The value of $\eta$ is within (0, 1]. If the sensing area of a node does not overlap with that of any other nodes, the value of $\eta$ is one. On the other hand, if the sensing area of multiple nodes overlaps the sensing area of a node, the value of $\eta$ may be much less than one.

For the whole sensor network, the sensing coverage is the ratio between the total sensing area covered by all sensor nodes and the target area. For example, if the target is 1000 $m^2$ and all of the sensors totally cover 900 $m^2$(due to some sensors failure), the sensing coverage of this sensor network now is 0.9.

## 3.4. Framework Design

The aim of the multi-agent framework is to improve the overall routing performance of existing routing approaches. In this framework, each agent can autonomously create a cooperative neighbour set and each agent can dynamically add or remove cooperative neighbours into or from its cooperative neighbour set. Afterwards, when a sensor has a packet to send, it sends the packet to one of its cooperative neighbours, and the cooperative neighbour then relays the packet to one of the cooperative neighbour's cooperative neighbours until the packet reaches the sink node. For the packet transmission between cooperative neighbours, any existing routing approach can be used.

### 3.4.1. Cooperative Neighbour Set Creation

The creation of a cooperative neighbour set is based on the sensors' historical information. We use an example to describe the creation procedure. A sensor, $v_j$, often relays packets for $v_i$. Then, $v_j$ may want itself to be $v_i$'s cooperative neighbour. $v_j$ then sends a request message to $v_i$ to suggest $v_i$ add $v_j$ as one of $v_i$'s cooperative neighbours. The request message includes the information regarding $v_j$ itself, such as the remaining energy of $v_j$, the normalised effective sensing area of $v_j$, the distance between $v_j$ to the sink node and the average number of hops from $v_i$ to $v_j$ (that is derived from the previously relayed packets, each of which contains a hop-count field). Once $v_i$ receives this request packet, $v_i$ will evaluate the reward of adding $v_j$ as one of $v_i$'s cooperative neighbours.$v_i$ estimates that if it adds $v_j$ as a cooperative neighbour, $v_i$'s storage consumption may increase and $v_j$'s energy consumption may rise (because in the future, $v_i$ will send more packets to $v_j$ than before). Such a consumption is a negative reward. On the other hand, however, in the future, the packet routing delay could be reduced, because the routing procedure becomes more targeted. Furthermore, the energy consumption of other sensors could be reduced, since packets would be relayed by fewer sensors. The storage and energy consumption can be estimated using Equations (1), (2) and (4). The estimation of packet routing delay is based on two factors: the communication radius of a sensor, $r$, and the distance between sensor node $v_i$ and sink node $s$, which is recorded as $d_{vi}{\rightarrow}s$. If there is more than one sink in the sensor network, such a distance demonstrates the shortest distance. In this study, it is assumed that each sensor has the same communication radius, $r$. The distance between sensor nodes to the sink node can be obtained during the initialisation stage of a sensor network. Once the two factors, $r$and $d_{vi}{\rightarrow}s$, are known, the estimation of packet routing delay can be obtained using Equation (6), where $\delta$ is a constant coefficient, which is greater than one.

$$del = \delta \cdot \frac{d_{v_i \rightarrow s}}{r} \tag{6}$$

The reason for using $r$ and $d_{vi}{\rightarrow}s$ to calculate the estimation of the packet routing delay is that the result of using $d_{vi}{\rightarrow}s$ to divide $r$ can approximately indicate the number of hops from sensor $v_i$ to sink $s$. Since a packet transmitted between two neighbouring sensors needs one time slot, the number of hops from sensor $v_i$ to sink $s$can approximately reflect the number of time slots needed to transmit a packet from $v_i$ to $s$, namely the packet delay from $v_i$ to $s$.

Hence, if $v_i$ wants to add $v_j$ as one of its cooperative neighbours, the estimated increase of storage consumption is $S^{(2)} \cdot \hat{t}_{ij}^{(3)}$, where $\hat{t}_{ij}^{(3)}$ indicates the intended time slots during which $v_i$ keeps $v_j$ as one of its cooperative neighbours; the estimated increase of energy consumption on $v_j$ is $\zeta \cdot Ene_{Rj}(k)$, where $k$ denotes the total size of messages (measured in bits) passed from $v_i$ to $v_j$ and $\zeta$ is a constant coefficient, which is greater than one; the estimated routing delay reduction is $\zeta \cdot (del_i - del_j)$; the estimated reduction of energy consumption on other sensor nodes is $\zeta \cdot hop_{i \to j} \cdot (Ene_T(k, \frac{r}{\delta}) + Ene_R(k))$. After the estimation, the reward of $v_i$, namely $\mathcal{R}_i$, is derived by summing up the four estimated values. Here, the four estimated values can be assigned different weights in different situations to demonstrate their different importance levels. For example, if in some situations, energy consumption is the most important, the energy consumption will be assigned the highest weight among these estimated values. In this paper, we do not consider special use cases. Thus, the four estimated values are assigned the same weight, *i.e.*, one.

Once the reward value, $\mathcal{R}_i$, is obtained, sensor $v_i$ has to make a decision regarding whether to add $v_j$ as one of its cooperative neighbours. A Q-learning algorithm is developed for this decision making process (see Algorithm 1). Q-learning is one of the most-studied reinforcement learning (RL) algorithms and has been applied with success in several domains, from relatively simple toy problems, such as cliff-walking [30], to more complex ones, such as web-based education [31] and face recognition [32]. One of the attractive features of Q-learning is that it assumes no knowledge about the environment (such as state transition functions or reward functions).

**Algorithm 1.** Learning progress of sensor $v_i$

Initialise $Q$ value for each available action arbitrarily;
**for** $k = 0$ to a predefined integer **do**;
  calculate $\pi$;
  **for each** available action $a \in A_i$ **do**;
    $Q_{k+1}(a) = Q_k(a) + \pi(a)\alpha_1(\sum_a \mathcal{R}(a)\pi(a) - Q_k(a))$;
  **end for**
**end for**
$a_{opti} \leftarrow argMax(Q)$;
$v_i$ takes the action $a_{opti}$;

Sensor $v_i$ first arbitrarily initialises the Q-value of each available action (Line 1). There are two available actions for sensor $v_i$: whether or not to add $v_j$ as one of its cooperative neighbours. Next, $v_i$ launches the learning process for the Q-value of each available action (Lines 2–7). In Line 2, the number of iterations is set to 30, which can yield adequate results. In Line 3, $\pi$ is calculated using Equation (7) for each available action, $a$, which is the $\epsilon$-greedy exploration method [33], where $0 < \varepsilon < 1$ is a small positive number and $n$ is the number of available actions of a sensor. The $\varepsilon$-greedy exploration method defines a semi-uniform probability distribution, in which the current best action is selected with probability $1 - \epsilon$ and a random action is chosen with probability $\epsilon$.

$$\pi(a) = \begin{cases} (1-\epsilon) + (\epsilon/n), \text{if } Q(a) \text{ is the highest value} \\ \epsilon/n \text{ , otherwise} \end{cases} \tag{7}$$

Line 5 demonstrates the Q-value update equation, where $0 < \alpha_1 < 1$ is the learning rate and $\mathcal{R}(a)$ is the estimated reward received by $v_i$ after taking action $a$. After finishing learning, $v_i$ executes the action that could maximise the Q-value (Lines 8–9). Finally, if $v_i$ decides to add $v_j$ as one of its cooperative neighbours, $v_i$ will send an acknowledgement message back to $v_j$ to let $v_j$ know that $v_j$ has been added to $v_i$'s cooperative neighbour set. Conversely, if $v_i$ does not decide to add $v_j$ as its cooperative neighbour, $v_i$ just ignores $v_j$'s request message and does not send any message back in order to save energy consumption. Having this algorithm, the expected behaviour of sensors can be derived by tracking the actions with the highest Q-value over the learning process of each sensor. The importance of obtaining a Q-learning algorithm with $\varepsilon$-greedy exploration is also justified through a large number of applications. For example, Galstyan *et al.* [34] applied a Q-learning algorithm with $\varepsilon$-greedy exploration to develop a decentralised resource allocation mechanism; Gomes and Kowalczyk [35] studied the problem of learning demand functions; and Ziogos *et al.* [36] investigated the development of bidding strategies.

Similarly, if $v_j$ has been a cooperative neighbour of $v_i$, but $v_j$ seldom relays packets for $v_i$, then $v_i$ may remove $v_j$ from its cooperative neighbour set so as to reduce $v_i$ computational overhead associated with taking $v_j$ into account whenever $v_i$ transfers a packet. The decision making of $v_i$, regarding removing $v_j$ from its cooperative neighbour set, is analogous to the decision making process about adding $v_j$ as a cooperative neighbour.

Of course, a sensor can directly choose an action that could maximise its reward, and thus, the learning process is not really necessary.

Kaelbling *et al.* [37] note, however, that this kind of algorithm, which always takes the highest rewards action and overlooks the trade-off between exploitation and exploration, may converge to a sub-optimal state.

### 3.4.2. Selecting a Cooperative Neighbour for Relaying

When a sensor, say $v_i$, has a packet to transfer, it needs to decide which cooperative neighbour is the best choice for relaying the packet. The selection of a cooperative neighbour is based on each cooperative neighbour's properties, which are embodied in the previous request message (As time progresses, the information embodied in the request message may be outdated. Thus, if a cooperative neighbour's status significantly changes, it will send a new message to $v_i$ regarding its current status.). These properties include the remaining energy of $v_j$, the normalised effective sensing area of $v_j$, the distance between $v_j$ to the sink node and the average number of hops from $v_i$ to $v_j$. Generally, sensor $v_i$ prefers to forward packets to a cooperative neighbour, which has more remaining energy, because such a cooperative neighbour is not likely to run out of energy. $v_i$ also prefers a small normalised effective sensing area, because even if such a cooperative neighbour is depleted, the overall coverage area will not be greatly affected, since the sensing area of that cooperative neighbour is overlapped by multiple sensors. A short distance from $v_j$ to the sink and a small number of hops from $v_i$ to $v_j$ are also preferred, as a short distance and a small number of hops usually imply little transmission delay. Therefore, for $v_i$, there are different benefits for choosing different cooperative neighbours to relay a packet. Such a benefit, for $v_i$ choosing $v_j$, can be calculated using Equation (8).

$$\mathcal{B}(v_j) = \gamma_1 \cdot Rem_E ng(v_j) - \gamma_2 \cdot \eta(v_j) - \gamma_3 \cdot d_{v_j \to s} - \gamma_4 \cdot hop_{i \to j}, \tag{8}$$

where $Rem_{E}ng(v_j)$ means the remaining energy of $v_j$, $\eta(v_j)$ indicates the normalised effective sensing area of $v_j$, $d_{vj \to s}$ is the distance from $v_j$ to the sink node, $hop_{i \to j}$ denotes the number of hops from $v_i$ to $v_j$ and $\gamma_1$, $\gamma_2$, $\gamma_3$, $\gamma_4$ are coefficients that are used to normalise the four values in the same magnitude.

After calculation of the benefit for selecting each cooperative neighbour, sensor $v_i$ employs another Q-learning algorithm to select a cooperative neighbour to forward the packet (Algorithm 2).

In Line 1, $v_i$ initialises the Q-value of each cooperative neighbour as zero. $\boldsymbol{\pi}$ is a vector, which demonstrates the probability with regard to selecting each cooperative neighbour to transfer the packet, $i.e., \boldsymbol{\pi} = \langle \pi(v_1), ..., \pi(v_{|N_i'|}) \rangle$. $v_i$ initially sets each element in $\pi$ to $\frac{1}{|N_i'|}$ for simplicity. Then, in Lines 2–4, $v_i$ updates the Q-value of each cooperative neighbour based on the benefit of selecting each cooperative neighbour, $\mathcal{B}$ ($v_j$), where $\alpha_2$ is the Q-learning rate. $v_i$ averages the benefits of selecting its cooperative neighbours and uses this average value and the newly created Q-values to update the probability vector $\pi$, where $\tau$ is the policy learning rate (Lines 5–8). The probability vector update method was developed by Zhang *et al.* [38]. In Line 9, the function *Normalise*() is used to constrain $\pi$ to a legal probability distribution, $i.e., \sum_{j=1}^{|N_i'|} \pi(v_j) = 1$, and *Normalise*($\pi$) has the minimum Euclidean distance to $\pi$. Finally, in Line 10, $v_i$ selects a cooperative neighbour to transfer the packet based on the probability vector $\pi$. This selection is akin to the roulette, namely that cooperative neighbour $v_j$ with a large value of $\pi$ ($v_j$) is more likely to be selected. Algorithm 2 is a direct policy search algorithm that learns stochastic policies. As argued in [39], stochastic policies can work better than deterministic policies in partially observable environments (such as packet routing in WSNs, where each sensor has only partial information about the entire WSN). In this algorithm, the policy search is based on the difference between the benefit of an action and the current average benefit. Thus, this algorithm encourages each sensor to select the cooperative neighbours, which are associated with high benefits, for packet relaying. Since the benefit of selecting each cooperative neighbour is dynamic, a sensor does not always select relaying neighbours from a small part of its cooperative neighbours. Therefore, the relaying role can be automatically rotated among the sensor's cooperative neighbours, which can then prolong the lifetime of each cooperative neighbour.

**Algorithm 2.** Choice of a cooperative neighbour for relaying

\\ according to $v_i$'s view \\
$v_i$ initialises Q-values and $\pi$;
**for each** cooperative neighbour of $v_i$, *i.e.*, $v_j \in N_i'$ **do**
  $Q(v_j) \leftarrow (1 - \alpha_2)Q(v_j) + \alpha_2 \cdot \mathcal{B}(v_j)$;
**end for**
$\bar{\mathcal{B}} \leftarrow \frac{1}{|N_i'|}\sum_{v_j \in N_i'} \mathcal{B}(v_j)$;
**for each** cooperative neighbour of $v_i$, *i.e.*, $v_j \in N_i'$ **do**
  $\pi(v_j) \leftarrow \pi(v_j) + \tau(Q(v_j) - \bar{\mathcal{B}})$;
**end for**
$\pi \leftarrow Normalise(\pi)$;
$v_i$ selects a cooperative neighbour based on $\pi$;

In this framework, each sensor node in the network has to have a global unique ID to distinguish different nodes in the network (see Figure 1). Although some addressing schemes, e.g., [40,41], can be used to handle this problem, it is still sometimes difficult for each node to have a global unique ID, especially in a large WSN. This is a drawback of this framework, and overcoming this drawback will be the subject of one of our future studies.

## 3.5. Complexity Analysis of the Framework

This framework is proposed as an assistant tool to enhance the performance of existing routing approaches, but meanwhile, the framework incurs extra complexity, which is independent of the complexities of specific underlying routing approaches. Such complexity is incurred during the creation and use of cooperative neighbour sets. As described above, a node, e.g., $v_i$, first estimates the reward of adding another node, e.g., $v_j$, to$v_i$'s cooperative neighbour set. Then, node $v_i$ uses Algorithm 1 to decide whether or not to add $v_j$ as one of its cooperative neighbours. In the future, when $v_i$ has a packet to send out, it will choose one of its cooperative neighbours to relay the packet.

The estimation of reward does not incur much extra complexity, as there is only the computation of simple equations (Equations (1)–(5)). Then, for the decision making of a node, in Algorithm 1, there is a nesting loop (Lines 2–7). The number of iterations of the loop, Lines 2–7, is 30. The number of iterations of the loop, Lines 4–6, is two, as there are two available actions: adding a node as a cooperative neighbour or not. Because the number of iterations of both of the loops is constant, the complexity of Algorithm 1 is $\mathcal{O}(1)$. For the selection of cooperative neighbours, in Algorithm 2, there are two loops. The number of iterations of each loop equals the number of cooperative neighbours of a sensor node. In the worst case, the number of cooperative neighbours of a sensor node is the number of sensor nodes in the WSN. As the number of sensor nodes in the WSN is $n$, the complexity of Algorithm 2 is $\mathcal{O}(n)$. It should be noted that to alleviate the potential worst case, sensor nodes are allowed to remove inefficient cooperative neighbours (recall Section 3.4.1).

In addition, there is also storage and communication consumption. Every time a node adds another node as a cooperative neighbour, the focal node has to use some storage space to store such a neighbouring relationship. In the worst case, a node adds all of the $n$ nodes in the WSN to be its cooperative neighbours. Thus, the storage complexity is $\mathcal{O}(n)$. Furthermore, before a node adds another node as a cooperative neighbour, a message has to be sent. Similarly, in the worst case, a node adds all of the $n$ nodes in the WSN to be its cooperative neighbours, which means that $n$ messages have to be sent. Thus, the communication complexity is $\mathcal{O}(n)$, as well. Certainly, the potential worst case can be alleviated by allowing sensor nodes to remove inefficient cooperative neighbours.

# 4. SIMULATION AND ANALYSIS

As illustrated in the previous section, the proposed framework is not a concrete routing approach, but rather a framework used to enhance existing routing approaches. Therefore, to assess the performance of the proposed framework, three existing routing approaches are selected, and each routing approach belongs to one of the three categories described in Section 2. Specifically, the selected routing approaches are gossiping [11] (a flat routing approach), LEACH-revised (a hierarchical routing approach) and GEAR [26] (a location-based routing approach). The reason for choosing the three routing approaches is that they are classical and easy to implement. Certainly, other routing approaches can also be employed in our simulation, but using the three classical approaches is sufficient to demonstrate the effectiveness of the proposed framework. Another of our future studies will take other more complex routing approaches into simulation. The three selected approaches are described in detail as follows.

1. Gossiping: In gossiping, a sensor node sends the packet to a randomly selected neighbour, which picks another random neighbour to forward the packet to, and so on, until the expiry time of the packet or the destination is reached, whichever comes first.

2. LEACH-revised: LEACH [19] randomly selects a pre-determined number of sensor nodes as cluster heads and rotates this role to evenly distribute the energy load among the sensors in the network. The role rotation is depicted as below. A sensor node chooses a random number, $r$, between zero and one. If this random number is less than a threshold value, $T(n)$, the node becomes a cluster head for the current round. The threshold value is calculated based on an equation that incorporates the desired percentage, $p$, to become a cluster head in the current round and the set of nodes that have not been selected as a cluster head in the last1p rounds. Then, all elected cluster heads broadcast an advertisement message to the rest of the nodes in the network that they are the new cluster heads. The problem with LEACH, however, is that it assumes that the cluster heads can directly send packets to the sink node, which is infeasible in large-scale WSNs. Thus, we introduce Yu *et al.*'s multi-hop routing algorithm [42] used by cluster heads to transfer packets to the sink node. Yu *et al.*'s algorithm is based on the cost of each wireless link, which is evaluated by using the remaining power of each node.

3. GEAR: There are two phases in GEAR. First, forwarding packets toward the target region: upon receiving a packet, a node relays the packet to the neighbour which is closer to the target region than itself. If there is more than one neighbour, the nearest neighbour to the target region is selected. However, if all of the neighbours are further from the destination region than the node itself, one of the neighbours is picked based on their evaluated cost to the destination region. Second, forwarding packets within the region: if the packet has arrived at the target region, it is diffused in the region by either recursive geographic forwarding or restricted flooding, until the packet reaches the sink node.

## 4.1. Simulation Setup

The simulation is divided into two parts. One is in a stationary WSN, and the other is in a dynamic WSN. In the stationary WSN, the position of each sensor node, including sink nodes, is fixed, meaning that sensors cannot move, and the WSN is closed, which means that no new sensors will join the network. On the other hand, in the dynamic WSN, each sensor node, involving sink nodes, has a pre-defined probability to move, and the WSN is open, which means that at each time slot, with a pre-determined probability, a new sensor may join the network. Performance is measured by the four quantitative metrics: average delivery latency, successful delivery ratio, number of live nodes and total sensing coverage. The meanings of the four metrics can be easily deduced from their names. For packet delivery, it is supposed that a packet is transmitted and then received in one time slot. The simulation is operated in a simulated 1000 m × 1000 m area, and 200 sensor nodes and five sink nodes are deployed randomly in this area. The communication radius, that is the furthest delivery radius, of each sensor is 60 m, and the sensing coverage radius of each sensor is 40 m. The initial energy of each sensor is set to 5 J. The average size of a packet is 50 bytes, and the actual size of a packet is based on the normal distribution with variance equal to 10. At each time slot, each node generates a packet based on a pre-defined probability, and the expiry time of a packet is based on exponential distributions. If a packet cannot reach a sink before its expiry time, the delivery of this packet is considered as a failure. Undelivered packets are also accounted for in the average delivery latency computation. The values and meanings of those parameters mentioned in the previous section, are listed in Table 1. These values were chosen experimentally to provide the best results. Each simulation was executed 200 times, and the average results are displayed in the following figures. We used JAVA to build the simulation platform, where two two-dimensional arrays were used to represent the two layers of a sensor network and another two-dimensional array was defined to indicate the distance between two neighbouring sensors. 'Sensor' was programmed as a class, and each sensor is an object of this class.

## 4.2. Simulation Results: The Stationary Occasion

Figure 2 demonstrates the performance of the three routing approaches and their multi-agent framework-based enhanced versions in a stationary WSN. The solid lines denote the performance variation of the three routing approaches, while the dashed lines indicate the performance of their multi-agent framework-based counterparts.

**Table 1.** Parameter settings.

| Parameters | Values | Explanations |
|---|---|---|
| $E_{elec}$ | $50\ nJ/bit$ | Energy dissipation parameter |
| $\epsilon_{amp}$ | $0.1\ nJ/bit/unit$ | Energy dissipation parameter |
| $\beta$ | 3 | Path loss exponent |
| $S^{(1)}$ | 0.01 | Storage coefficient for keeping packets |
| $S^{(2)}$ | 0.0015 | Storage coefficient for keeping neighbours |
| $\delta$ | 2.5 | Path length coefficient |
| $\zeta$ | 2.2 | Energy consumption coefficient |
| $\alpha_1, \alpha_2, \tau$ | 0.1, 0.3, 0.2 | Learning rate |
| $\epsilon$ | 0.4 | Action selection distribution probability |
| $k$ | 20 | Learning rounds |
| $\gamma_1, \gamma_2, \gamma_3, \gamma_4$ | 1.7, 8.2, 0.013, 1.5 | Benefit coefficients |

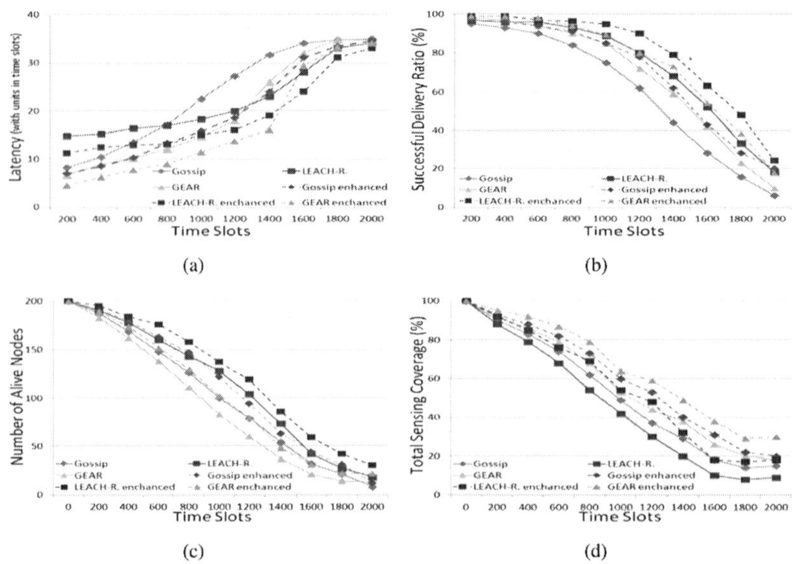


**Figure 2.** Performance of routing approaches in a stationary WSN. (**a**) Average delivery latency; (**b**) successful delivery ration (%); (**c**) number of live nodes; (**d**) total sensing coverage (%).

In Figure 2a, initially, the delivery latency values in gossiping and GEAR are less than that in LEACH-revised. However, as time progresses, LEACH-revised turns out to be better than both gossiping and GEAR. This can be explained as follows. Initially, in LEACH-revised, the cluster heads usually have to wait for a period for their cluster members to send packets to them and then aggregate and condense these packets to save transmission energy consumption. As time progresses, some sensor nodes are depleted. In gossiping and GEAR, some pre-existing routing paths will disappear, but in LEACH-revised, the impact, caused by sensor node failure, is much less than that in gossiping and GEAR, because routing in LEACH-revised is based on only a few nodes, *i.e.*, cluster heads, and even if these cluster heads are exhausted, some other nodes can take their roles immediately by taking the measure developed in LEACH.

In Figure 2b, LEACH-revised achieved the highest successful delivery ratio. This is due to the fact that packets in LEACH-revised use the fewest hops from a source node to the sink node compared to gossiping and GEAR, because, as described above, routing in LEACH-revised is based only on cluster heads. Thus, it is more reliable for LEACH-revised to successfully transmit a packet to the sink node than both gossiping and GEAR.

In Figure 2c,d, it can be found that in gossiping and LEACH-revised, more sensor nodes can be saved than that in gossiping (Figure 2c), but in GEAR, the total sensing coverage is larger than that in both gossiping and LEACH-revised (Figure 2d). This finding confirms the argument in [21] that saving energy consumption can prolong the lifetime of sensors, but does not necessarily achieve better sensing coverage. This can be explained by the fact that GEAR's transmission method can easily deplete the nodes that are closer to the sink node, but gossiping's transmission method, randomly choosing a neighbour for transmission, can distribute the energy consumption over a larger area than GEAR's method does; so, in gossiping, fewer nodes are depleted compared to GEAR in the same time span. Likewise, in LEACH-revised, routing only relies on cluster heads, and the role of a cluster head is rotated among different sensor nodes. Thus, the lifetime of each sensor node in LEACH-revised could be further prolonged. However, due to the same reason, the total sensing coverage in GEAR is larger than that in gossiping, because gossiping's transmission method consumes energy over a larger area, which implies that depleted sensors are distributed over a larger area; so, the total sensing coverage reduction is more than that in GEAR. Similarly, in LEACH-revised, the energy dissipation in the network is distributed more broadly than it is in gossiping, because in LEACH-revised, sensors are organised based on clusters that are broadly distributed over the network. In this way, when sensors in different clusters are depleted, the total sensing coverage will be decreased more than that in gossiping. Finally, it can be seen that the performance of the three routing approaches is improved by the use of the proposed framework, measured by the four quantitative metrics. This is due to the fact that the framework enables each sensor to find the most useful cooperative neighbours, and therefore, the routing is more targeted. In addition, as described in the previous section, in the framework, each sensor takes into account energy consumption, storage consumption and sensing coverage, when it builds its cooperative neighbour set. Thus, by using the framework, the performance of the three existing routing approaches can be raised in all four quantitative metrics.

## 4.3. Simulation Results: The Dynamic Occasion

In this simulation, at each time slot, each sensor has the probability, 0.05, to move from its current position to a random position, and at each time slot, a new sensor will arrive at the network with the probability 0.2 at a random position. When an existing sensor moves to a new position or a new sensor joins the network, its physical neighbour set and cooperative neighbour set will be initialised. The node's physical neighbour set is built based on its communication radius. This means that any node in this node's communication radius will become this node's physical neighbours. The node's cooperative neighbour set is initialised in the same way as the physical neighbour set. In practice, the movement of a sensor is usually based on some purpose, so the sensor does not move to a random position. In this simulation, the aim is just to demonstrate how the proposed framework works in a dynamic WSN, so we simplified the movement of a sensor by setting its movement destination to a random position.

In Figure 3, the main trend of the simulation results is similar to that in a stationary WSN, but the overall performance of the three routing approaches and their multi-agent framework-based enhanced versions is improved compared to that in a stationary WSN. This is because sensors movement can, to some extent, balance the energy consumption in the network. In a stationary WSN, those sensors that are neighbours of the sink nodes are very likely to be depleted, because they are the final hops to the sink nodes and all of the packets have to be relayed by them. Therefore, if sensors can move, those final hop nodes will most likely move away and, likewise, other sensors will most likely be the final hop nodes; so, the energy consumption can be balanced to some extent. Analogously, if new sensors can join the network, the energy of the whole network increases, which is good for the network in various aspects. There are still some specific phenomena that should be noted. In Figure 3a, the average delivery latency in LEACH-revised in the final time slots is longer (*i.e.*, worse) than that in gossiping and GEAR, which is different from Figure 2a. This is ascribed to the fact that routing in LEACH-revised is based on cluster heads, so new sensors joining or sensor movement cannot effectively shrink the routing path, and thus, the latency of packet routing cannot be improved very much. Due to the same reason, the difference of the final performance, regarding successful delivery ratio (Figure 3b) between LEACH-revised and gossiping is reduced compared to Figure 2b. On the other hand, in Figure 3c, the difference in the number of live nodes in LEACH-revised and gossiping in the final time slots is enlarged compared to Figure 2c, which can reflect the fact that LEACH-revised is quite good at balancing energy-consumption. Finally, again, using the proposed framework can improve the performance of all three routing approaches in a dynamic WSN.

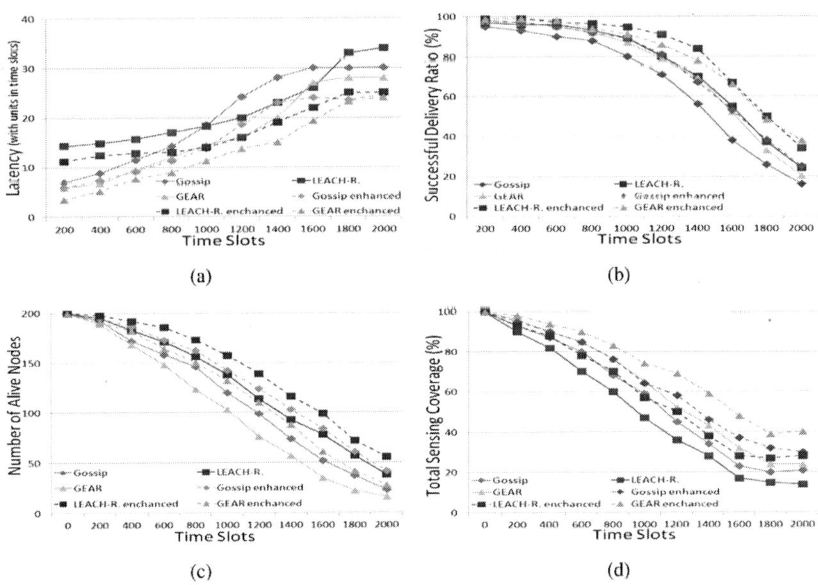


**Figure 3.** Performance of routing approaches in a dynamic WSN. (**a**) Average delivery latency; (**b**) successful delivery ration (%); (**c**) number of live nodes; (**d**) total sensing coverage (%).

In summary, using the proposed framework, the performance of the selected three routing approaches in both stationary and dynamic WNSs can be improved by around 10%–15% with regard to the four quantitative metrics, which demonstrates the potential usefulness of this framework.

## 5. CONCLUSIONS AND FUTURE WORK

In this paper, a multi-agent framework was developed, which to the best of our knowledge, is the first one used to assist many existing routing approaches to improve their routing performance. In addition, this framework is decentralised and does not need global information. The simulation results exhibited the effectiveness of this framework in both static and open and dynamic WSNs.

In the future, we first intend to overcome the drawback that an addressing scheme has to be used. Another flaw is that this framework uses a 'unit disk' communication model, where sensors can communicate bi-directionally if they are in each other's communication radius. However, in real-world WSNs, several physical links are uni-directional. Additionally, more existing routing approaches have to be taken into our simulation to further validate the effectiveness of our framework, and it will be very interesting to allow nodes to use different routing approaches in different situations. We are also interested in validating the framework in heterogeneous WSNs, where different sensors have different sensing coverage and initial energy. Finally, when such studies are finished, we intend to test this framework on a specific platform, e.g., NS2 [43], and further in a real environment.

## ACKNOWLEDGEMENTS

This work was originally created at University of Wollongong, Wollongong 2522, Australia. It was partly supported by an ARC Discovery Project (DP150101775) from Australian Research Council, Australia. We also greatly appreciate Mrs. Madeleine Strong Cincotta, who provided professional language editing for us.

## AUTHOR CONTRIBUTIONS

Dayong Ye and Minjie Zhang created the major part of this paper. Yun Yang contributed a lot during the revision process by providing complexity analysis.

## REFERENCES

1. Winkler, M.; Tuchs, K.D.; Hughes, K.; Barclay, G. Theoretical and practical aspects of military wireless sensor networks. *J. Telecommun. Inf. Technol.* **2008**, *2*, 37–45.

2. Seema, A.; Reisslein, M. Towards efficient wireless video sensor networks: A survey of existing node architectures and proposal for a flexi-wvsnp design. *IEEE Commun. Surv. Tutor.* **2011**, *13*, 462–486.

3. Tavli, B.; Bicakci, K.; Zilan, R.; Barcelo-Ordinas, J.M. A survey of visual sensor network platforms.*Multimed. Tools Appl.* **2012**, *60*, 689–726.

4. Akkaya, K.; Younis, M. A survey on routing protocols for wireless sensor networks. *Ad Hoc Netw.* **2005**, *3*, 325–349.

5. Li, C.; Zhang, H.; Hao, B.; Li, J. A Survey on Routing Protocols for Large-Scale Wireless Sensor Networks.*Sensors* **2011**, *11*, 3498–3526.

6. Anisi, M.H.; Abdullah, A.H.; Razak, S.A.; Ngadi, M.A. An Overview of Data Routing Approaches for Wireless Sensor Networks. *Sensors* **2012**, *12*, 3964–3996.

7. Al-Karaki, J.N.; Kamal, A.E. Routing Techniques in Wireless Sensor Networks: A Survey. *IEEE Wirel. Commun.* **2004**, *11*, 6–28.

8. Sohrabi, K.; Gao, J.; Ailawadhi, V.; Pottie, G.J. Protocols for Self-organization of a Wireless Sensor Network.*IEEE Pers. Commun.* **2000**, *7*, 16–27.

9. Polastre, J.; Hui, J.; Levis, P.; Zhao, J.; Culler, D.; Shenker, S.; Stoica, I. A unifying link abstraction for wireless sensor networks. Proceedings of the Third ACM Conference on Embedded Networked Sensor Systems, San Diego, CA, USA, 2–4 November 2005; pp. 76–89.

10. Tsai, M.; Yang, H.; Liu, B.; Huang, W. Virtual-Coordinate-Based Delivery-Guaranteed Routing Protocol in Wireless Sensor Networks. *IEEE/ACM Trans. Netw.* **2009**, *17*, 1228–1241.

11. Hedetniemi, S.; Liestman, A. A Survey of gossiping and broadcasting in communication networks.*Networks* **1988**, *18*, 319–349.

12. Heinzelman, W.; Kulik, J.; Balakrishnan, H. Adaptive protocols for information dissemination in wireless sensor networks. Proceedings of the 5th Annual ACM/IEEE International Conference on Mobile Computing and Networking, Seattle, WA, USA, 15–19 August 1999; pp. 174–185.

13. Kulik, J.; Heinzelman, W.R.; Balakrishnan, H. Negotiation-Based Protocols for Disseminating Information in Wireless Sensor Networks. *Wirel. Netw.* **2002**, *8*, 169–185.

14. Rogers, A.; David, E.; Jennings, N.R. Self-organized routing for wireless microsensor networks. *IEEE Trans. Syst. Man Cybern. Part A Syst. Hum.* **2005**, *35*, 349–359.

15. Beraldi, R.; Baldoni, R.; Prakash, R. A biased random walk routing protocol for wireless sensor networks: The lukewarm potato protocol. *IEEE Trans. Mob. Comput.* **2010**, *9*, 1649–1661.

16. Quang, P.T.A.; Kim, D.S. Enhancing Real-Time Delivery of Gradient Routing for Industrial Wireless Sensor Networks. *IEEE Trans. Ind. Electron.* **2012**, *8*, 61–68.

17. Li, Y.; Chen, C.S.; Song, Y.Q.; Wang, Z.; Sun, Y. Enhancing realtime delivery in wireless sensor networks with two-hop information. *IEEE Trans. Ind. Electron.* **2009**, *5*, 113–122.

18. Niu, J.; Cheng, L.; Gu, Y.; Shu, L.; Das, S.K. R3E: Reliable Reactive Routing Enhancement for Wireless Sensor Networks. *IEEE Trans. Ind. Inform.* **2014**, *10*, 784–794.

19. Heinzelman, W.; Chandrakasan, A.; Balakrishnan, H. Enerty efficient communication protocol for wireless sensor networks. Proceedings of the 33rd Hawaii International Conference System Sciences, Maui, HI, USA, 4–7 January 2000; pp. 3005–3014.

20. Lindsey, S.; Raghavendra, C. PEGASIS: Power-Efficient Gathering in Sensor Information Systems. *IEEE Aerosp. Conf. Proc.* **2002**, *3*, 1125–1130.

21. Tsai, Y.R. Coverage-Preserving Routing Protocol for Randomly Distributed Wireless Sensor Networks.*IEEE Trans. Wirel. Commun.* **2007**, *6*, 1240–1245.

22. Valentini, G.; Abbas, C.J.B.; Villalba, L.J.G.; Astorga, L. Dynamic Multi-Objective Routing Algorithm: A Multi-Objective Routing Algorithm for the Simple Hybrid Routing Protocol on Wireless Sensor Networks.*IET Commun.* **2010**, *4*, 1732–1741.

23. Villalba, L.J.G.; Orozco, A.L.S.; Cabrera, A.T.; Abbas, C.J.B. Routing protocols in wireless sensor networks.*Sensors* **2009**, *9*, 8399–8421.

24. Mottola, L.; Picco, G.P. MUSTER: Adaptive Energy-Aware Multisink Routing in Wireless Sensor Networks.*IEEE Trans. Mob. Comput.* **2011**, *10*, 1694–1709.

25. Xu, Y.; Heidemann, J.; Estrin, D. Geography-informed energy conservation for ad hoc routing. Proceedings of the 7th Annual ACM/IEEE International Conference on Mobile Computing and Networking, Rome, Italy, 16–21 July 2001; pp. 70–84.

26. Yu, Y.; Estrin, D.; Govindan, R. *Geographical and Energy-Aware Routing: A Recursive Data Dissemination Protocol for Wireless Sensor Networks; Technical Report*; UCLA Computer Science Department: Los Angeles, CA, USA, 2001.

27. Huang, H.; Hu, G.; Yu, F.; Zhang, Z. Energy-aware interference-sensitive geographic routing in wireless sensor networks. *IET Commun.* **2011**, *5*, 2692–2702.

28. Awad, A.; German, R.; Dressler, F. Exploiting Virtual Coordinates for Improved Routing Performance in Sensor Networks. *IEEE Trans. Mob. Comput.* **2011**, *10*, 1214–1226.

29. Petrioli, C.; Nati, M.; Casari, P.; Zorzi, M.; Basagni, S. ALBA-R: Load-Balancing Geographic Routing Around Connectivity Holes in Wireless Sensor Networks. *IEEE Trans. Parallel Distrib. Syst.* **2014**, *25*, 529–539.

30. Sutton, R.S.; Barto, A.G. *Reinforcement Learning: An Introduction*; MIT Press: Cambridge, MA, USA, 1998.

31. Iglesias, A.; Martnez, P.; Aler, R.; Fernndez, F. Learning teaching strategies in an Adaptive and Intelligent Educational System through Reinforcement Learning. *Artif. Intell.* **2009**, *31*, 89–106.

32. Harandi, M.T.; Ahmadabadi, M.N.; Araabi, B.N. Optimal Local Basis: A Reinforcement Learning Approach for Face Recognition. *Int.J. Comput. Vis.* **2008**, *81*, 191–204.

33. Gomes, E.R.; Kowalczyk, R. Dynamic Analysis of Multiagent Q-Learning with E-Greedy Exploration. Proceedings of the ICML'09, Montreal, QC, Canada, June 2009.

34. Galstyan, A.; Czajkowski, K.; Lerman, K. Resource allocation in the grid using reinforcement learning. Proceedings of the AAMAS'04, New York, NY, USA, July 2004; pp. 1314–1315.

35. Gomes, E.R.; Kowalczyk, R. Learning the IPA Market with Individual and Social Rewards. Proceedings of the International Conference on Intelligent Agent Technology (IAT'07), Fremont, CA, USA, 2–5 November 2007; pp. 328–334.

36. Ziogos, N.P.; Tellidou, A.C.; Gountis, V.P.; Bakirtzis, A.G. A reinforcement learning algorithm for market participants in FTR auctions. Proceedings of the 7th IEEE Power Tech Conference, Lausanne, Switzerland, 1–5 July 2007; pp. 943–948.

37. Kaelbling, L.P.; Littman, M.L.; Moore, A.W. Reinforcement Learning: A Survey. *J. AI Res.* **1996**, *4*, 237–285.

38. Zhang, C.; Lesser, V.; Shenoy, P. Multi-Agent Learning Approach to Online Distributed Resource Allocation. Proceedings of the IJCAI'09, Pasadena, CA, USA, July 2009; pp. 361–366.

39. Singh, S.P.; Jaakkola, T.; Littman, M.L.; Szepesvari, C. Convergence Results for Single-Step On-Policy Reinforcement Learning Algorithms. *Mach. Learn* **2000**, *38*, 287–308.

40. Akyildiz, I.F.; Su, W.; Sankarasubramaniam, Y.; Cayirci, E. Wireless sensor networks: A survey. *Comput. Netw.* **2002**, *38*, 393–422.

41. Shah, R.; Rabaey, J.M. Energy Aware Routing for Low Energy Ad Hoc Sensor Networks. *Proc. IEEE Wirel. Commun. Netw. Conf.* **2002**, *1*, 350–355.

42. Yu, M.; Leung, K.K.; Malvankar, A. A Dynamic Clustering and Energy Efficient Routing Technique for Sensor Neworks. *IEEE Trans. Wirel. Commun.* **2007**, *6*, 3069–3079.

43. The Network Simulator. Available online: http:// www.isi.edu/ nsnam/ ns/ (accessed on 28 April 2015).

# 10

# CHAPTER

# INTELLIGENT MULTI-AGENT BASED INFORMATION MANAGEMENT METHODS TO DIRECT COMPLEX INDUSTRIAL SYSTEMS

**Danilo Avola[1], Luigi Cinque[2], Giuseppe Placidi[1]**

[1]*Department of Life, Health and Environmental Sciences, University of L'Aquila, L'Aquila, Italy*

[2]*Department of Computer Science, Sapienza University of Rome, Rome, Italy*

## ABSTRACT

In recent years, the increasingly complexity of the logistic and technical aspects of the novel manufacturing environments, as well as the need to increase the performance and safety characteristics of the related cooperation, coordination and control mechanisms is encouraging the development of new information management strategies to direct and manage the automated systems involved in the manufacturing processes. The Computational Intelligent (CI) approaches seem to provide an effective support to the challenges posed by the next generation industrial systems. In particular, the Intelligent Agents (IAs) and the Multi-Agent Systems (MASs) paradigms seem to provide the best suitable solutions. Autonomy, flexibility and adaptability of the agent-based technology are the key points to manage both automated and information processes of any industrial system. The paper describes the main features of the IAs and MASs and how their technology can be adapted to support the current and next generation advanced industrial systems. Moreover, a study of how a MAS is utilized within a productive process is depicted.

**Keywords:** Industrial Systems; Information Management; Intelligent Agents; Multi-Agent Systems

## 1. INTRODUCTION

The current industrial systems are destined to become environments even more technologically advanced in every aspect. This is due to several reasons: the complexity of the new manufacturing processes, the need to optimize costs, the increasing of the complexity in managing resources, the need to improve performances, the aspects tied to the information exchanges, the manufacture of complex products, the need to ensure advanced safety standard, and so on. All these aspects imply a continuous and growing renovation of the current industrial activities. Artificial Intelligence (AI) techniques [1,2], utilized to overcome these new qualitative and quantitative challenges bring from the next industrial age, seem to provide the more suitable support since they naturally have all these technical features (e.g., adaptability, reasoning, learning) required from the new manufacturing processes [3]. In particular, the use of AI based strategies allow to the industrial environments to adopt autonomous, dynamic and intelligent procedures to face expected and unexpected issues. The IAs and MASs are technologies that can play an important role in every aspect involved in the development of the next generation of the advanced industrial systems. In fact, these systems have to be designed to include intelligent, autonomous and evolvable entities in turn composed by different sub-entities having the same features. The number of levels (i.e., entities, sub-entities) and the complexity of each entity and sub-entity depend on the specific kind of the industrial system. The mentioned features are strongly tied to the IAs and MASs paradigms [4,5]. Moreover, several IAs and MASs have been just adopted in a wide range of applications that are connected to advanced industrial systems, such as: Robotic, artificial vision, production planning, systems control, engineering, information exchanging, and so on. In what follows, we describe the main features of the IAs and MASs and how their technology can be adapted to support the current and next generation advanced industrial systems.

The paper is organized as follows. Section 2 introduces the general aspects of the logical architecture of the current industrial systems. Section 3 describes the main features of the IAs and MASs to be adapted to support industrial systems. Section 4 provides a study of how a MAS is utilized within a productive process. In particular, the section discusses the different MASs algorithms highlighting their use to support the real-time production process. Section 5 shows the principles of the hybrid systems and presents the main characteristics of a case study. Section 6 concludes the paper.

## 2. CURRENT INDUSTRIAL SYSTEMS ARCHITECTURE

The logical architectures of the current industrial systems can be seen as a correlated set of connected crucial processes that have to achieve a prefixed objective taking into account the current elaboration state of every other process. All the processes are aimed to reach a common target: The Artifact. In our context, the last mentioned term has to be considered both a physical object (e.g., technical or mechanical device, industrial product) and an abstract object (e.g., general service, immaterial product). The concepts regarding the creation or delivery of an artifact can be generalized to provide a common description of current industrial system architectures.

Each process deals with a specific activity (or part of it) of the whole industrial process. In fact, a generic Industrial system includes heterogeneous activities, such as: economical planning, quality and test control, real-time production control, internal activities, processes monitoring, robot control, sales planning, and so on. For this reason, the specification of a process can be completely different from another. As shown in **Figure 1**, a process is an autonomous entity characterized by internal rules and protocols. Every process works in coordination with others by taking into account their internal state.

The internal rules and protocols are highly dependent from the specific process. In general, the output of a process (i.e., artifact and information) regards a collection of homogeneous activities, while the input (i.e., processes interaction) concerns the connection of several heterogeneous activities, such as: Technical and schedulling information, control and manufacturing activities, data, artifact (or part of it). Every single process is composed of several different tasks (or sub-tasks) concerning a specific part of the whole industrial process. As shown in **Figure 2**, a task (or a sub-task) is an activity that can be solved, by decisional processes, using autonomous and intelligent methods without a specific knowledge of the surrounding processes. The depth of the tree depends on both the kind of process and its complexity. Usually, processes concerning long term economic planning and technical activities are more complex than others.

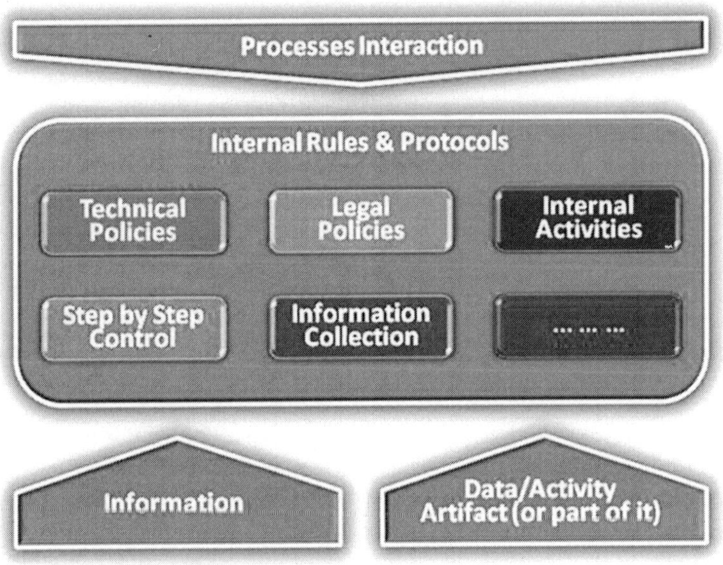


**Figure 1**. A general scheme of a process.

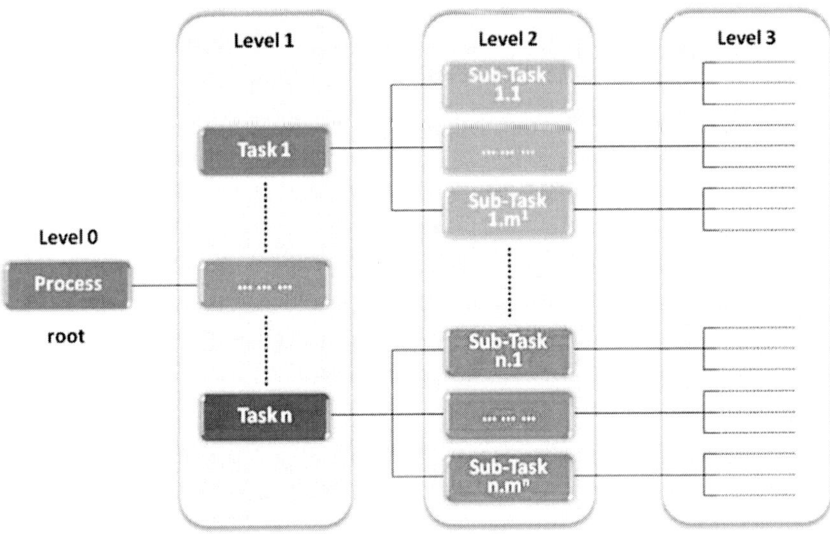


**Figure 2**. Process, tasks and sub-tasks hierarchy.

The process at the root level (i.e., level 0) provides each of its tasks the needed input to accomplish the related assignments. Likewise, each further task (i.e., level 1) provides each of its sub-tasks the suitable part of the original input to accomplish the related sub-assignments. This activity of splitting input and assignments ends at the last level of the tree. The solution of each task and sub-task, by a decisional process, can require information or activities coming from different tasks and/or sub-tasks connected to any tree level. For this reason, the interaction mechanisms are a critical aspect of the current industrial systems. A failure in a task can produce a critical crash of the related process, or of the connected processes. Furthermore, the last mentioned aspect is fundamental to establish the intelligent mechanisms to solve the heterogeneous activities belonging to the complex industrial systems (e.g., real-time monitoring, dynamic business planning, chain control, simulation environments) and to overcome the expected and unexpected issues (e.g., breakdowns, environment changes, damages).

The decisional process that solves a task at the level 1 of the hierarchy can interact with any other decisional process that solves any other task at the same hierarchical level. Subsequently, the decisional processes that solve sub-tasks on the others levels of the hierarchy (e.g., level 2) have to directly interact only with the decisional processes that solve a sub-task connected at the same task or sub-task at the previous hierarchical level.

In others worlds, on each tree level, only the "brothers" can directly interact. If decisional processes belonging to different "fathers" have to interact, then the same "fathers" will accomplish the interaction activity.

As shown in **Figure 3**, the artifact is the result of joined processes. Likewise the previously described cooperation strategy, each decisional process that solves a process within the level 0 can interact with any other decisional process that solves any other process at the same level 0. Also in this case, if two decisional processes, driving two different tasks (e.g., at level 1) belonging to two different processes, have to interact, the same two processes will accomplish the interaction activity. This kind of hierarchical interaction serves to avoid disorder (i.e., anarchy) between tasks, sub-tasks and processes. Moreover, it is useful to allow the decisional process to have a total autonomy with respect the fixed assignment.

Each decisional process involved in the solution of a process, task or sub-task has to have well defined features. These features specify the way (e.g., protocols of communication and execution, priority of the decisional tree, priority of the queues, relationships between tasks or sub-tasks) and the resources (e.g., technical, economical, time planning, supply chain management) to carry on the activities. A generic decisional process can be considered as a set of activities that can be carried on by an intelligent, autonomous and flexible entity. Technically, these entities can be seen as a set of interactive procedures able to learn and evolve according to their life cycle. For these reasons, the use of IA and MAS technologies are considered the best solutions to these kinds of issues. As shown in **Figure 4**, each decisional process should be driven by a single IA. Each IA knows only the rules, the procedures and the protocols related to its specific task. It has a limited point of view regarding the whole industrial process and it is able to adopt solutions only on its specific contextual environment. Moreover, the networked set of the IAs (i.e., MASs) should drive each process into the industrial system. This association between tasks, assignments and intelligent entities allows to implement independent customized solutions to solve global and local specific issues.

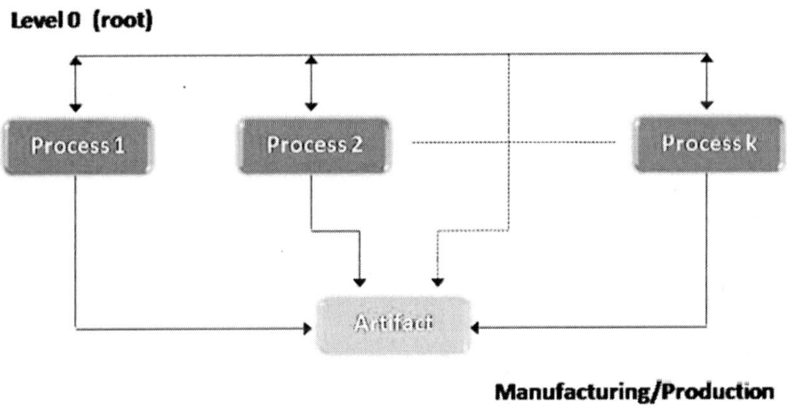


**Figure 3**. Level 0: The processes.

# 3. IAS AND MASS IN INDUSTRIAL SYSTEMS

The agent and multi-agent technologies have been widely adopted in robotic and automatic control system fields. The growing complexity of the current industrial environments and their affinity to the mentioned fields has encouraged the design of IAs and MASs aimed to direct each aspect of the industrial chain. The CI based techniques promise to be the best way to perform these historical changes. There are several similar definitions of intelligent agent [6,7]. Considering the set of these definitions, in our context an intelligent agent can be described as: A social, autonomous and intelligent computational entity capable to accomplish a definite objective through an adaptive reasoning. Regardless of how an IA is defined, it is characterized by several general features [8] that can be adapted to our context. These features, joined with the given definition, complete the IA paradigm.

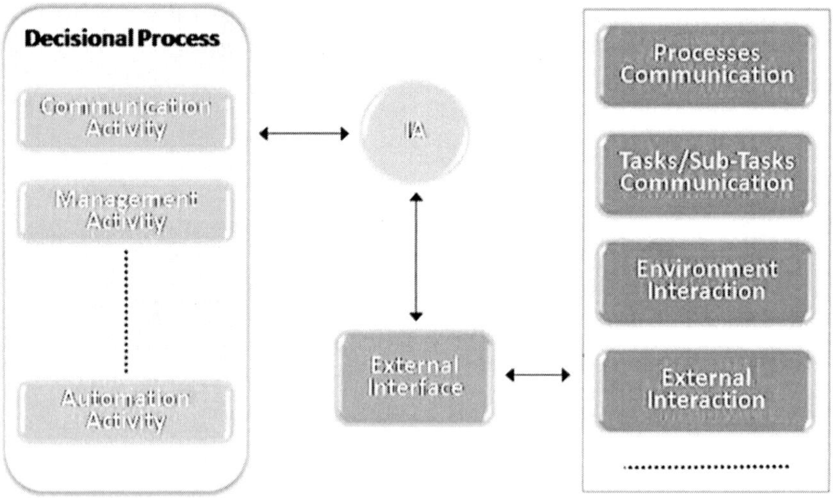


**Figure 4**. Decisional process and intelligent agent.

## 3.1. IA Properties Adapted to Industrial Systems

There are five main properties belonging to the IA that have to be adapted to build a basic industrial oriented IA.

### 3.1.1. Reactivity

An IA is a reactive entity that continuously perceives and affects with its surrounding environment. In an industrial system this feature depends on the process on which the agent is applied. For example, in economical planning processes the reactivity can be considered a not active feature, i.e., the agent has a starting receptive state and it will tend to switch its condition only if it recognizes particular changes in the environment.

Instead, in technical processes the reactivity can be considered active, i.e., each change in the environment is considered an important event that has to be analyzed in real-time.

### 3.1.2. Pro-Activity

An IA has to operate in an expected way, but it can also take initiatives to overcome unexpected events. Usually, this feature does not depend on the process, it is tied to the environment in which the agent has to operate. The environments into the industrial processes can be classified in deterministic and non-deterministic. At the first class often belong processes derived from economical tasks that are tied to several unexpected changes (e.g., economical plans, supply chain management). At the last class almost always belong processes derived from technical tasks that are usually scheduled and coordinated (e.g., test and control, robot interaction, real-time monitoring).

### 3.1.3. Autonomy

An IA has to be capable of autonomous actions to accomplish its objectives. There are several situations in which an agent has to autonomously decide what is necessary to do, and how it has to be done. The main issue of this feature is to decide which, how many and when specified actions/activities can be adopted by the IA. As general rule, an IA is authorized to take autonomous decisions only if there is the real possibility that the whole process crashes. These situations occur quite often in advanced industrial systems where several technological actors are involved. In fact, the introduction of complex systems in the industrial environments (e.g., real-time systems, on-line robot vision monitoring, automatic control management) has led high level of entropy. Usually, a reasoning decision tree is used to decide the actions that have to be performed [9,10]. By the autonomy property each agent can interact with the environments or other agents without an external presence to ensure the achievement of the prefixed objectives.

### 3.1.4. Flexibility

An IA has to interact with the surrounding environment in different ways. It has to have the capability to adopt quickly itself to drastic changes into the relationships, environments or events. Sometimes conventional communication or interaction ways could be not sufficient to reach a prefixed objective. For example, an agent could want data from another agent in different way respect the standard protocols. For this reason, an agent has to manage its characteristics (e.g., output, external interface, internal protocols). This property is particularly important in industrial systems because they are prone to dynamic changes in management and technical levels.

### 3.1.5. Social Ability

An IA has to interact with other agents (also humans). This feature is the core of the intelligent agent theory; through the interaction an agent can understand the events and adapt its characteristics to the dynamic situations surrounding it. An IA needs to communicate with the internal and external environments to ensure several main activities (e.g., find out the state of other agents, sub-tasks synchronize, tasks scheduling, acknowledgement). Furthermore, it has to be considered that some agents (e.g., mediator agents, contractor agents, negotiator agents) have particular assignments that, more than others, need to interact with the surrounding entities. In industrial systems this feature takes a key value in business plans and technical activities.

The last property introduces a main matter in the industrial systems, i.e., the languages through which the agents exploit their own social ability. In general, the most important challenges in Agent Oriented Software Engineering (AOSE) is to use or create a suitable Agent Oriented Programming Language (AOPL) to allow the agents an effective and efficient interaction. There are several advanced tools that aid the developers in designing complex agent based software systems. Moreover, there are several powerful programming languages designed to facilitate the programming of agent entities. The chosen of the correct tool and language to achieve a prefixed assignment is a critical matter. Independently from the agent implementation there are some issues (e.g., debugging, validation, verification) that depend on the specific context in which the agents have to operate.

In AOPLs the focus is on how to describe the behavior of an agent in terms of constructs [11] (e.g., plans, communication strategies, interaction patterns, goals, messages). This agent description is aimed to specify the "reasoning state" of the agent during its activities. A typical example is AGENT-0 [12] a simple and powerful generic agent interpreter. This AOPL tends to program agents in terms of basic behavior, in this way the properties of the agents can be developed by definite notions, such as: Reactions, communication ways, interactions, rules, and so on. Another interesting AOPL is dMARS (Distributed Multi-Agent Reasoning System) [13], which provides both a sophisticated monitor and manageable macros to develop interaction activity between agents. Moreover, this AOPL provides several functionalities to manage critical implementation steps (e.g., configurations of internal states, perception of events). The dMARS can be considered a reference technology to direct complex industrial systems since it has a wide range of high solutions that support the whole industrial processes. A last interesting AOPL considered to develop agents in industrial systems is JACK [14]. It is a versatile framework designed to create, by models, the concepts that drive the intelligent agents. This language has a wide use in industrial contexts since it provides powerful tools for the run-time support. Moreover, JACK has predefined structured patterns to manage critical situations (e.g., concurrency, reaction to events, failure). In the last years, there have been many efforts to support the IA communication in different fields, including the Industrial systems, the result has been a growing development of multi-purpose languages (e.g., 2APL, 3APL) [15]. An agent has several others features that can be considered implicitly included in the previous features (e.g., learn capability, reasoning, auto-improvement).

In complex industrial systems there are several kinds of agents according to different processes or tasks. For example, the communication-agents are specialized in communication-activities between sets of agents. They deal with strategies and protocols for the information or data exchange. Another example is the learning-agents that learn about the functionality and potentiality of the surrounding environment and transmit this knowledge to the other agents. Several others dedicated agents are developed according to specific requirements. Beyond the potentiality of a single IA the algorithms that drive the industrial processes are multi-agent based.

MASs can be considered as composed by heterogeneous agents, which are aimed to achieve common objectives. Likewise to the intelligent agent definition, also in this case there are several similar definitions of MAS [16]. In our context a MAS can be described as a suitable and reasoning assemble of agents that, according to their features, achieve common objectives through connections and teamwork intelligent mechanisms. Also the MASs have the same features observed for the intelligent agents: the main difference is the coordination and cooperation processes that drive the agents. Moreover, a MAS has interaction properties that define the collective reasoning of the whole system. Usually, each MAS has a set of rules and protocols that are inherited from each single agent. These last define the kind of collective intelligence characterizing each single MAS. These considerations conclude the paradigm definition of MASs.

A complex MAS can be composed by different sets of MASs obtaining a system with high levels of communication processes. In any case, these systems are characterized from more complex coordination, cooperation and teamwork mechanisms. Usually, no centralized control methods are implemented, but specific strict rules and hierarchic behavior strategies are performed to ensure the correct working of the system. The difficulty in the MASs management depends on the number of agents and their complexity. **Figure 5** shows a typical MAS scheme where each process is driven by a set of agents that have different dedicated assignments.

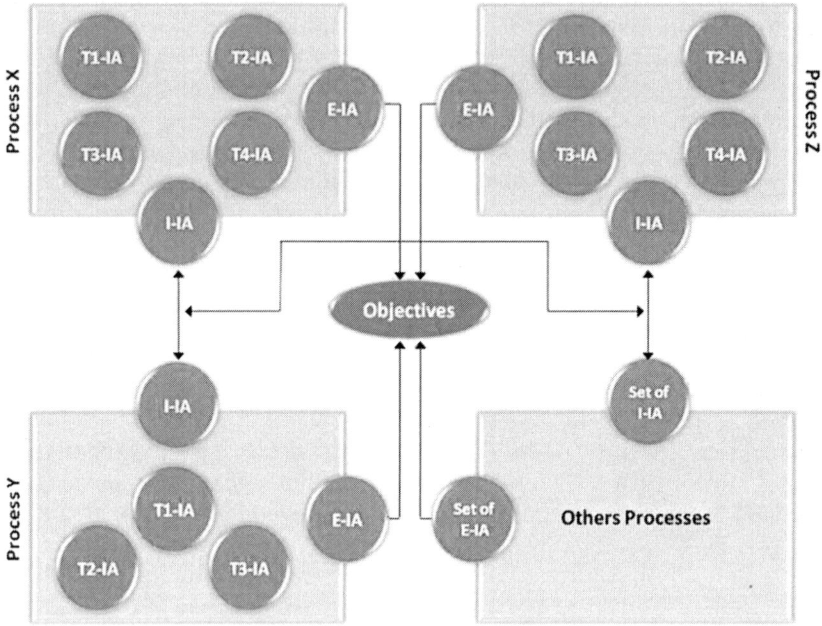


**Figure 5**. Simple MAS scheme.

Each process (X, Y, Z and others) is composed by some tasks that are accomplished by several agents. As presented in **Figure 5**, the process X is composed by four industrial activities (T1, T2, T3 and T4, where T = Task), and two service activities (E and I, where E = Executor, I = Interface). All the processes involved in the industrial environment, independently from their role, are aimed to reach the common target: the objective (artifact/activity).

The agents tied to the industrial activities accomplish specific tasks tied to the industrial process, while the others two activities provide coordination and cooperation mechanisms. I-IA and E-IA work on the whole process. In fact, the first provides the result of the Industrial process (e.g., artifact or part of it, data, information) to the environment; the second uses a communication channel to interact with other processes (i.e., with the related I-IAs). Note that only the I-IA and E-IA can make actions outside their own process. The other four agents (T1-IA, T2-IA, T3-IA and T4-IA) can accomplish only internal assignments and can interact only with agents belonging to the same process.

# 4. MASS: APPROCHES AND ALGORITHMS

The new generation of the industrial systems can be conceived like Distributed Artificial Intelligence (DAI) systems described as cooperative systems where a set of heterogeneous agents act jointly to solve a given problem [17]. A DAI system belongs to the Distributed Problem Solving (DPS) field where a problem is solved by using several modules [18] which cooperate in dividing and sharing knowledge about the common issue. All these factors influence the choice of the multi-agent architecture to drive the algorithms used into each agent and the related communication strategies. There are several multiagent architectures used to perform the activities of a large set of heterogeneous environments; the most effective are inspired by the distributed version of BDI (Belief-Desire-Intention) [19]. This architecture is based on the correspondence between beliefs, desires and intentions with reciprocal identifiable data structures. The BDI based systems have several features (e.g., agent scalability, real-time communication, acknowledge mechanisms). They can be managed to be adapted in any environment. In **Figure 6** a general overview of the basic patterns to implement industrial architectures is given. They arc based on the classical configurations used in others fields, the differences are came out during the implementation by considering the features of each agent in accordance to the specific industrial system.

As a general rule, each task is assigned to a specific Task Master Agent (TMA) which, according to the specific architecture, can solve different assignments (e.g., safety and security, execution, coordination, cooperation). The TMA can work alone or supported by several subagents, depending on the nature and complexity of the task. These sub-agents deal with only a peculiar and homogeneous step of the task. Moreover, they are usually execution-oriented (i.e., executor agents) or information-oriented (i.e., acknowledger agents). Commonly, the sub-agents can have relationships only with sub-agents of the same task or with the related TMA. A particular kind of TMA, the Gateway Task Master Agent (GTMA), has to interact with all the TMAs belonging to a same process. Finally, the Process Master Agent (PMA) is used to allow interaction of different agents. Each GTMA can interact only with the related PMA, each PMA can interact with any other PMA. The communication activities are usually performed by well-known protocols and algorithms in the network field. Indeed, each communication, in the MASs, tends to be implemented according to FIPA specifications [20]. It is the IEEE Computer Society organization that promotes agent-based technology and the interoperability of its standards with other technologies.

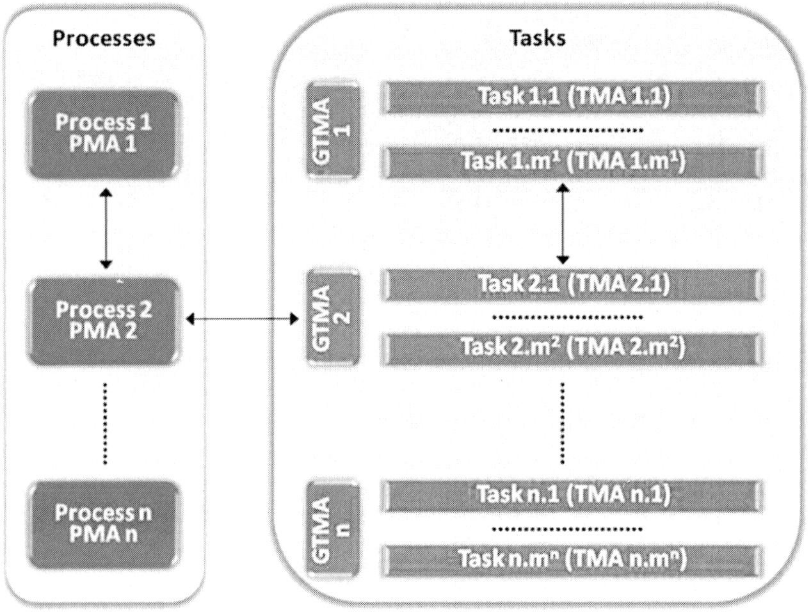


**Figure 6**. Sample of architectural industrial system.

The MASs communication activity can be analyzed by graph theory [21], where each node is an intelligent agent and the edges between nodes are the communication channels. This allows to adopt the well-known algorithms to manage environments composed by multi-agent systems. The current MASs communication processes are inspired to the basic algorithms on network architectures [22]: Routing, broadcasting, and semi-group computation.

The routing algorithms regard the processes used to visit a graph to reach a particular objective. In literature exists different algorithms to accomplish this task [23]. For example, the Dijkstra and Bellman-Ford algorithms research a shortest-path from a single source node to any other node in the graph. A variant of these algorithms is given by Floyd-Warshall and Johnson that provide strategies to research a shortest-path from multiple source nodes. Also the Prim and Kruskal algorithms (minimum spamming tree) and the Ford-Fulkerson and Karp algorithms (maximum flow) are commonly used in routing issues. All these algorithms are used to allow the agents different communication strategies which are tied to several factors, such as: type and aim of the interaction, kind of activity, kind of involved agents, and so on.

The broadcasting algorithms provide a direct way to communicate to a node in a graph. These algorithms are based on one-to-all philosophy, where a single node (i.e., agent) has to send (i.e., interact) a message to all the nodes of the graph. The weighs in the graph can drive the communication process according to the specific objective. They are usually represented by vectors that include a wide set of information (e.g., priority access value, identification value (ID), agent state value). In industrial contexts these algorithms are used to support control activities and to provide commands and scheduling steps to technical processes.

The semi-group computation algorithms are based on the binary communication of the intelligent agents composing a definite and closed set of agents. In general, the binary communication is used to reach each node in the graph through a dynamic bridge that allows two agents to communicate according to some convenience rule (e.g., distance between the agents, cost of the connection, priority level). These algorithms are usually applied, in industrial environments, where a strict intelligent agent structure is required. The particular structure of a semi-group of agent is suitable to implement systems able to react to non-deterministic events (e.g. external operations, unexpected broken events, sabotages). In addition to the three mentioned basic algorithms (and their evolutions) on network architectures, there are several experience-based approaches for everyday use which provide ad-hoc solutions allowing a qualitative improvement of the management standards. In this way, the human experience can be utilized to optimize the agent implementation improving the industrial process.

**Table 1** presents a brief description of the communication algorithms according to the different main industrial environments. The physical connection of a MAS is a crucial point of the current and next generation Industrial environments, in fact the way in which sets of agents are connected each other can drive the chooses about the adopted interaction communication technologies. For example, in large-scale management processes centralized and hierarchical architectures are preferred. On the contrary, in advanced manufacturing processes are used flexible, scalable and modular architectures. **Figure 7** shows that the architectures for industrial systems are implemented with four approaches: Star, ring, chain and network.

**Table 1**. Communication algorithms.

| Industrial Environments | Algorithms | | |
|---|---|---|---|
| | **Routing** | **Broadcasting** | **Semi-Group** |
| **Management and Strategies** | general spamming tree | one-to-all one-to-set one-to-one And vice versa | binary computation |
| **Supply Chain Management** | maximum flow shortest path | one-to-all set-to-set and vice versa | binary computation linear computation |
| **Control and Driving Tools** | minimum spamming tree maximum spamming tree | one-to-all one-to-set and vice versa | binary computation multiple computation |
| **Security Activities** | multiple shortest path general spamming tree | one-to-one one-to-all and vice versa | multiple computation |

The star-structure is used in processes, tasks and subtasks where a centralized mechanism is needed. The activities are managed by the central agent that can directly interact with each node in the graph. A typical example of this architecture regards the interaction between the interface intelligent agent (i.e., I-IA, the central node) and the other agents belonging to the same task. A different approach is provided by the ring-structure where a decentralized interaction mechanism is given. The responsibilities and the activities are subdivided between the agents involved in the graph. A typical example is given in the processes managed from a large amount of heterogeneous agents. The chain-structure provides a hierarchical interaction structure. This approach is commonly used to allow the communication between different levels. This interaction is the basic communication approach used in the industrial systems where every level has to solve different assignments. Each node can interact only with its own father or sons. The network-structure provides an interaction mechanism where each node can directly interact with any other. In this architecture the nodes have similar assignments. It is usually used to allow the agent of a same level to interact each other.

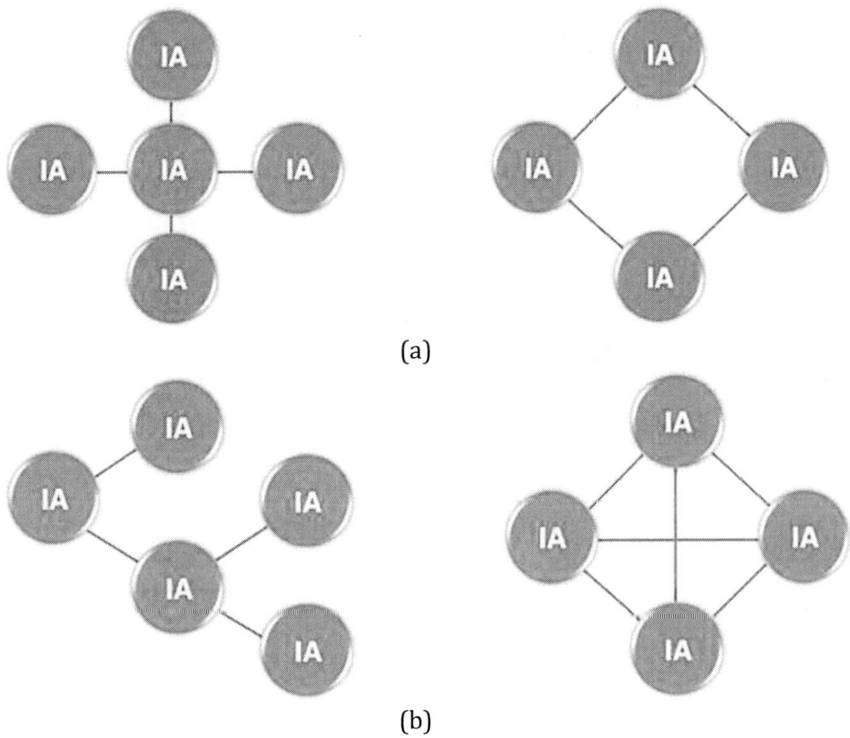


**Figure 7**. (a) Star; (b) ring; (c) chain; (d) network structures.

In our context, an industrial process can be classified within one of the following classes: Planning and Strategies of Management Processes (PSMP), Driving and Control of Manufacturing Processes (DCMP), Advanced Analysis of Critical Processes (AACP). The PSMP term highlights the high level activities that drive the whole business processes. These activities cover each level of the industrial architecture and include a wide range of assignments (e.g., economical strategies and planning, coordination of internal and external activities, definition of policies and rules). A classical application of the MASs regards the Supply Chain Management (SCM) [24]. A supply chain is a network that deals with several activities (e.g., materials, services, logistic planning) to reach the distribution of a final artifact. The PSMP are usually directed by statistical and computational algorithms. Each agent in the system controls a specific algorithm. The aim of the agent is to elaborate the information about the environments to obtain numerical values and, then, computes the values through statistical models to obtain a numerical transposition of the real world involved in the coordination activities. An interesting kind of algorithms regards the deduction/cognition process. Usually, these algorithms are performed by a mathematical-agent that knows linear programming and polynomial approximation techniques, useful to derive deductions. Other classes of algorithms are available according to the specific process.

Other important algorithms included in this processes class regard those to solve conflicts. The agents that accomplish this kind of algorithms exploit the social ability feature. In particular, fuzzy logic approaches [25] are implemented into the agent to face different issues (e.g., reliability of events, flexibility, mediator activity). **Table 2** shows a brief description of the common algorithms used in PSMP class.

The DCMP class regards all the activities concerning the manufacturing processes. This class deals with assignments, such as: Robots collective control, industrial tools management, technical safety policies, and so on. A classical application of this class regards the manufacturing robots control [26,27]. These algorithms regard the real-time reactivity on runtime environments. In fact, they must guarantee an immediate and solving interacttion with the surrounding context. Indeed, in this class it is possible to distinguish two sub-classes of algorithms. The first regards the algorithms inside the industrial tools (e.g., firmware or machine level), the second is about the algorithms inside the technological structures of the manufacturing environments (e.g., workstation, data processing centre). The algorithms belonging to the first sub-class are designed to be independent and light. These algorithms do not have a high computational level, but they have more parallel mechanisms to avoid and overcome bad events (e.g., failures, accidents). The most important agent is the interface-agent that controls and manages the interaction between the tool and the external world. In the second case the implemented algorithms are statistic-based. In particular, a large amount of control algorithms, and related agents, have knowledge about the discrete and continue random variables. In this way, for example, a prediction about a lot of targets can be done (e.g., life of a component, the state of an entity, the safety state of a tool). An interesting aspect related to the second sub-class of algorithms regards the simulation environment for manufacturing activities. These algorithms are used in modeling and simulation environments for improving or testing the different levels involved in the manufacturing activities. The algorithms used in this context regard the model understanding area (e.g., probabilistic models, prevision models). These algorithms are used to check the planning and scheduling strategies for a specific operative task. Moreover, they allow to the user to achieve simulations and predictions and to obtain different information about the events that occur during the plan strategy [28]. Also in this case **Table 3**shows a brief description of the main DCMP algorithms.

The last class of processes, AACP, is referred to critical tasks, sub-tasks and processes. This class considers each level of the industrial system, from technical up to economical issues. An interesting example is provided by the monitoring and diagnosis systems applied to sophisticate manufacturing environments (e.g., production of chips, production of complex artifacts, production of components, model testing). In this case the system provides several run-time dependent functionalities that perform activities as monitoring and diagnosis. These functionalities are accomplished by advanced algorithms (e.g., sensor understanding, pattern recognition, machine learning). If an agent has some doubts regarding the correct execution of a task, it may ask to interrupt each activity of the process. For example, particular agents can be created to recognize particular features on images that represent welding on an artifact in control process. The agents work on the images and they give information about the quality of the welding to the industrial manufacturing system. **Table 4** presents the main algorithms belonging to the AACP class.

**Table 2**. PSMP algorithms.

| PSMP | Algorithms |
|---|---|
| **Economical Planning** | deduction and reasoning, simulation, exhaustive research, linear approximations |
| **Supply Chain Management** | combinatory, linear programming, polynomial approximation, fuzzy logic |
| **Strategic Planning** | prediction, heuristic, comparison, numerical evaluation and synthesis |
| **Stock Management** | binary research, dictionary management, balanced trees, fuzzy logic |

# 5. HYBRID SYSTEMS: A CASE STUDY

An interesting opportunity regards the use of more than one computational intelligent (CI) approach to solve issues in industrial systems. For example, in [29] is described a DNA-MAS genetic programming system, which is designed for application-generic multi-agent simulation and generation using an advanced symbolic language built specifically for the use in a random mutation and crossover environment. Indeed, in industrial systems, the other techniques of artificial intelligence (e.g., genetic algorithms, neural networks) are used to improve specific features of an agent. For example, in a manufacturing environment, where the visual images of the artifact are important for the quality control, an agent could need to exploit the potentiality of a genetic algorithm (which approximates a set of solution to the optimal solution) to try a better identification of the meaningful features of the same images. There are several examples, not only in industrial systems, where hybrid architectures can result suitable to face critical problems. The purpose of this brief section is to highlight that the artificial intelligence techniques have a common root (i.e., the intelligent dynamic reasoning) that can be applied using different techniques to adopt a different representation of the human reasoning for solving activity.

**Table 3**. DCMP algorithms.

| DCMP | Algorithms |
|---|---|
| **Driving** | real-time tracking models, interactive patterns, experience-based models |
| **Control** | discrete and continue random variables, prediction models, reasoning-based models |
| **Runtime** | real-time acknowledgment, scheduling of dynamic queues |
| **Simulation** | probabilistic models, prevision models, non-deterministic models |

**Table 4**. AACP algorithms.

| AACP | Algorithms |
|---|---|
| **Control** | pattern recognition, deforming models, feature extraction, global and local operators |
| **Position** | tracking, scheduling, machine learning, feedback ranks, feedback management |
| **Driving** | feedback management, retroactivity strategies, experience-based models |
| **Simulation** | prediction, stochastic computation |

An interesting case study, shown in [30], regards a hierarchically organized multi-agent system for production control of semiconductor wafer manufacturing facilities (wafer fabrication). The production control of wafer fabrication is challenging from a complexity and coordination point of view. The goal of semiconductor manufacturing is the production of integrated circuits. This environment is characterized by several industrial tools that are managed, in different way, from a large amount of users. Moreover, the semiconductor manufacturing domain is characterized by stochastic events like machine breakdowns and the change of customer related due dates of the lots. A flexible representation of the process conditions of semiconductor manufacturing domain has been defined with the purpose to develop the MAS architectture. Then, it has been provided modeling capabilities for agent hierarchies; moreover it has been provided to the system the capabilities to emulate a wafer fabrication represented by a discrete event simulation model for performance assessment of our agent-based production control system. Finally, it has been given a mechanism about integration capabilities for legacy software to use more advanced heuristics for staff agents. Through this implemented strategy an efficient complex industrial system has been developed.

# 6. CONCLUSION

The evolution of the current industrial systems and the challenges of the next generation ones are encouraging the development of new information management strategies to direct and manage the automated systems involved in the manufacturing processes. The technology tied to the CI approaches, like IAs and MASs, seem to provide a valid solution to this kind of issues. This paper presents an overview of how efficient, autonomous and intelligent entities can support a new point of view on each and every aspect regarding a complex industrial environment.

## REFERENCES

1. R. Lee, "Computer and Information Science, Studies in Computational Intelligence," Springer-Verlag, Berlin and Heidelberg, 2012.

2. L. Benyoucef and B. Grabot, "Artificial Intelligence Techniques for Networked Manufacturing Enterprises Management," Springer-Verlag, London, 2010.

3. D. Laha and P. Mandal, "Handbook of Computational Intelligence in Manufacturing and Production Management," IGI Publishing, Hershey, 2007.

4. K. Hermann, "Distributed Manufacturing: Paradigm, Concepts, Solutions and Examples," Springer-Verlag, London, 2010.

5. W. Xiang and H. P. Lee, "Ant Colony Intelligence in MultiAgent Dynamic Manufacturing Scheduling," Journal of Engineering Applications of Artificial Intelligence, Vol. 21, No. 1, 2008, pp. 73-85.

6. D. S. Kim, C. S. Kim and K. W. Rim, "Modeling and Design of Intelligent Agent System," International Journal of Control, Automation, and Systems, Vol. 1, No. 2, 2003, pp. 257-261.

7. V. Graudina and J. Grundspenkis, "Technologies and MultiAgent System Architectures for Transportation and Logistics Support: An Overview," Proceedings of the International Conference on Computer Systems and Technologies (CompSysTech'05), Varna, 16-17 June 2005, pp. IIIA.6-1-IIIA.6-6.

8. S. J. Russell and P. Norvig, "Artificial Intelligence: A Modern Approach," 3rd Edition, Prentice-Hall Inc., New Jersey, 2010.

9. C. Z. Janikow, "Fuzzy Decision Trees: Issues and Methods," IEEE Transactions on Systems, Man, and Cybernetics, Part B: Cybernetics, Vol. 28, No. 14, 1998, pp. 1- 14.

10. P. Stone and M. Veloso, "Using Decision Tree Confidence Factors for Multi-Agent Control," Proceeding of the Second International Conference on Autonomous Agents (AGENTS'98), Minneapolis, 10-13 May 1998, pp. 86-91.

11. D. Mitrović, M. Ivanović and M. Vidaković, "Introducing ALAS: A Novel Agent-Oriented Programming Language," Proceeding of the International Conference on Numerical Analysis and Applied Mathematics, 2011, pp. 861-864.

12. Y. Shoham, "Agent-Oriented Programming," Journal of Artificial Intelligence, Vol. 60, No. 1, 1993, pp. 51-92.

13. M. D'Inverno, M. Luck, M. Georgeff, D. Kinny and M. Wooldridge, "The dMARS Architecture: A Specifiction of the Distributed Multi-Agent Reasoning System," Journal of Autonomous Agents and Multi-Agent Systems, Vol. 9, No. 1-2, 2004, pp. 5-53.

14. M. Winikoff, "JACK Intelligent Agents: An Industrial Strength Platform, Multi-Agent Programming: Languages, Platforms and Applications," Proceeding of the International Conference in Multiagent Systems, Artificial Societies, and Simulated Organizations, Vol. 15, No. 2, 2005, pp. 175-193.

15. K. V. Hindriks, F. S. De Boer, W. Van Der Hoek and J.-J. C. Meyer, "Agent Programming in 3APL," Proceeding of Autonomous Agents and Multi-Agent Systems, Vol. 2, No. 4, 1999, pp. 357-401.

16. H. Sevay and C. Tsatsoulis, "Multiagent Reactive Plan Application Learning in Dynamic Environments," Proceedings of the First International Joint Conference on Autonomous Agents and Multiagent Systems: Part 2 (AAMAS'02), Bologna, 15-19 July 2002, pp. 839-840.

17. N. M. Avouris and L. Gasser, "Distributed Artificial Intelligence: Theory and Praxis," Kluwer Academic Publishers, Norwell, 1991.

18. S. Kovalchuk, A. Larchenko and A. Boukhanovsky, "Knowledge-Based Resource Management for Distributed Problem Solving," Proceeding of the International Conference on Knowledge Engineering and Management, Vol. 123, No. 2012, 2012, pp. 121-128.

19. N. Ronald, "Modelling Pedestrian Behaviour Using the BDI Architecture," Proceeding of the IEEE/WIC/ACM International conference on Intelligent Agent Technology, Compiègne, 19-22 September 2005, pp. 161-164.

20. M.-P. Huget, "The Foundation for Intelligent Physical Agents," 2012. http://www.fipa.org

21. R. J. Trudeau, "Introduction to Graph Theory," 2nd Edition, Dover Publications, Mineola, New York, 1993.

22. B. Parhami, "Introduction to Parallel Processing: Algorithms and Architecture," Plenum Press, New York, 1999.

23. T. H. Cormen, C. E. Leiserson, R. L. Rivest and C. Stein, "Introduction to Algorithms," McGraw-Hill, New York, 2002.

24. T. Mentzer, "Supply Chain Management," Sage Publications Ltd., Thousand Oaks, 2000.

25. M. Mohammadian, "Designing Unsupervised Hierarchical Fuzzy Logic Systems," Machine Learning: Concepts, Methodologies, Tools and Applications, IGI Global, Hershey, 2011, pp. 253-261.

26. H. Colestock, "Industrial Robotics," McGraw-Hill, New York, 2005.

27. D. Patrick and S. Fardo, "Industrial Process Control Systems," 2nd Edition, The Fairmont Press, Lilburn, 2011.

28. S. Kraus, "Strategic Negotiation in Multiagent Environments, Intelligent Robotics and Autonomous Agents," The MIT Press, Cambridge, 2001.

29. S. J. Kollmansberger and S. L. Mabry, "Intelligent Agent Generation with the DNA-MAS Genetic Programming System," Proceeding of the International Conference on Artificial Intelligence and Soft Computing, Banf, 17-19 July 2002, pp. 101-116.

30. L. Monch, S. Marcel and J. Zimmermann, "FABMAS: An Agent-Based System for Production Control of Semiconductor Manufacturing Processes," Proceeding of the 1st International Conference on Industrial Applications of Holonic and Multi-agent Systems (HoloMAS'03), Prague, 1-3 September 2003, pp. 258-267.

# 11

# CHAPTER

# TIMED-AUTOMATA BASED MODEL-CHECKING OF A MULTI-AGENT SYSTEM: A CASE STUDY

**Nadeem Akhtar, Muhammad Nauman**

*Department of Computer Science and IT, The Islamia University of Bahawalpur, Baghdad-ul-Jadeed Campus, Bahawalpur, Pakistan*

## ABSTRACT

A multi-agent based transport system is modeled by timed automata model extended with clock variables. The correctness properties of safety and liveness of this model are verified by timed automata based UPPAAL. Agents have a degree of control on their own actions, have their own threads of control, and under some circumstances they are also able to take decisions. Therefore they are autonomous. The multi-agent system is modeled as a network of timed automata based agents supported by clock variables. The representation of agent requirements based on mathematics is helpful in precise and unambiguous specifications, thereby ensuring correctness. This formal representation of requirements provides a way for logical reasoning about the artifacts produced. We can be systematic and precise in assessing correctness by rigorously specifying the functional requirements.

**Keywords:** Software correctness, Formal verification, Model checking, Timed-automata, Multi-Agent System, Timed Computation Tree Logic (TCTL)

# 1. INTRODUCTION

The use of formal methods to define the requirements, design and architecture of a multi-agent system results in precise and unambiguous specifications. These formal specifications provide the basis for systematic, mathematically-proven, well-defined, and unambiguous software development phases of analysis, design, and implementation. Multi-agent systems are distributed, decentralized, consisting of autonomous computing entities known as agents. Correctness is a mathematical property that establishes the equivalence between software and its specifications. Software systems analyzed, designed, and implemented by using agents to offer significant challenges in ensuring their correctness. One of the methods of ensuring the correctness of safety and liveness properties of these agent-based systems is to use formal model checking methods based on timed automata.

# 2. STATE OF THE ART

## 2.1. Formal methods

The primary objective of a formal approach is precise and unambiguous specification. A representation of the requirements based on mathematics aids in precise specification of the software, thereby ensuring that the correctness, completeness, and unambiguous properties of the system are preserved [1] . The formal representation of software requirements provides a method for logical reasoning about the artifacts produced. This achieves more precision in the description and allows for a stronger design that satisfies the required properties. Formal methods offer the ability to rigorously prove system correctness, i.e. that specifications are consistent with the stated objective; that code is consistent with specification; and that code produces a desired result and none other. To overcome the complexity problems in multi-agent systems and get significant results with formal specifications, we must cope with complexity at each phase: requirements, architecture to design and implementation. We can prove the correctness of a multi-agent system by formalizing critical components in the multi-agent development life-cycle.

The most important reasons to use formal methods in software engineering are: rigorous analysis of system properties; property-preserving transformations; error-free implementation; high quality of each phase of the development process; firm foundation for the adaptation and evolution process; continuous correctness as multi-agent systems are concurrent and often have dynamic environments; formal specification and modeling of a multi-agent system architecture which can change at run-time (i.e. dynamic architecture); specification according to the functional and non-functional properties; property-preserving step-by-step transformations from abstract to concrete concepts and then stepwise refinement to implementation code; improved documentation and understanding of specifications.

## 2.2. Agents and Multi-Agent Systems

An agent is a computer system that is capable of autonomous actions on behalf of its user or owner [2]. Agents are coarse-grained computational systems, each making use of significant computational resources [3]. An agent is a software entity that is able to conduct information-related tasks without human supervision [4]. Intelligent agents (i.e. referred to as rational agents) are systems that accomplish their goals by acting rationally. Rational agents can use reasoning to make decisions about their goals. A rational agent is an autonomous computing entity that can accomplish tasks autonomously on the behalf of its user. It can also interact and collaborate with other agents to accomplish its goals. It can also refuse an order or an action that is called from another agent. Intelligent agents must show some degree of autonomy, social ability, and combine proactive and reactive behavior.

- Autonomous: agents have a degree of control on their own actions, they own their thread of control and under some circumstances, they are also able to take decisions;

- Proactive: agents do not only react in response to external events (i.e. a remote method call), but they also exhibit a goal-directed behavior and where appropriate are able to take initiative;
- Social: agents are able to and need to interact with other agents in order to accomplish their task and achieve the complete goal of the system [5].

The environment is a first-class abstraction that provides the surrounding conditions for agents to exist and that mediates both the interaction among agents and the access to resources [6]. A generic environment program has been defined by [7]. This simple program gives the agents precepts and receives back their actions; it then updates the state of the environment based on the actions of the agents and other dynamic processes in the environment that are not considered to be agents. Demazeau [8] considers four essential building blocks for agent systems: agents (i.e., the processing entities), interactions (i.e., the elements for structuring internal interactions between entities), organizations (i.e., elements for structuring sets of entities within the multi-agent system), and finally the environment that is defined as the domain-dependent elements for structuring external interactions between entities. The environment is an independent component of the multi-agent system that has its own responsibilities. These responsibilities are not dependent on agents. It provides the medium for agents to exchange messages, and all agent interactions are done through it.

A multi-agent system is composed with autonomous entities (i.e. agents) that interact with one another. Multiple agents are necessary to solve a problem, especially when the problem involves distributed data, knowledge, or control [9]. A multi-agent system is a collection of several interacting agents in which each agent has incomplete information or capabilities for solving the problem [10]. In a multi-agent system, agents are autonomous. There is no global system control, data is decentralized, and communication is asynchronous.

## 2.3. Correctness: Safety and liveness properties

A program is functionally correct if it behaves according to its stated functional requirements. Correctness is a mathematical property that establishes the equivalence between software and its specifications [11]. We can be systematic and precise in assessing correctness by rigorously specifying the functional requirements. Software systems provide critical services to users, i.e. process control systems in nuclear power plants or in chemical industry, radiation machines in hospitals, transport systems such as cars, trains and airplanes. In these types of systems, correctness is of vital importance.

Safety and liveness properties are correctness properties. The safety property is an invariant which asserts that something bad never happens, that an acceptable state of affairs is maintained. Magee and Kramer [12] have defined safety property S = {a1, a2 ... an} as a deterministic process that asserts that any trace including actions in the alphabet of S is accepted by S. ERROR conditions are like exceptions which state what is not required, as in complex systems we specify safety properties by directly stating what is required. The liveness property asserts that something good happens, which describes the states of a system that an agent must bring about given certain conditions. These properties play a vital role in system verification. Both safety and liveness properties are complementary to each other, safety alone or liveness alone is not sufficient to ensure system correctness. Progress is a type of liveness property. Progress P = {a1, a2 ... an} defines a property P which asserts that in an infinite execution of a target system, at least one of the actions (a1, a2 ... an) will be executed infinitely [13]. We have the safety and liveness properties mathematically based on timed automata and are unambiguous.

## 2.4. Formal verification

Formal verification is the mathematical demonstration of the correctness of a system. The basic idea is to construct a mathematical model of the system under investigation, a model which represents the possible behavior of the system. The correctness requirements are specified along with the other functional requirements that represent the desirable behavior of the system. Based on these specifications, we check formal proof whether the possible behavior agrees with the desired behavior. Verification process can be made precise by using formal methods. Formal verification leads to proving or disproving the correctness with respect to this formal correctness notion. Formal verification can achieve complete exhaustive coverage of the system thus ensuring that undetected failures in the behavior are excluded.

In summary, formal verification requires a model of the system consisting of:

1) A set of states incorporating information about values of variables program counters;
2) A transition relation that describes how the system can change from one state to another;
3) A specification method for expressing requirements in a formal way;
4) A set of proof rules to determine whether the model satisfies the stated requirements.

To obtain a more concrete feeling of what is meant, we consider the way in which sequential programs can be formally verified.

## 2.5. Model Checking

Model checking [14] - [19] is a method for automatic and algorithmic verification of finite state concurrent systems. It takes as input a finite state model of a system and a logical property, it then systematically checks whether this property holds for a given initial state in that model. Model checking is performed as an exhaustive state space search that is guaranteed to terminate since the model is finite. It uses temporal logic to specify correct system behavior. An efficient, flexible search procedure is used to find correct temporal patterns in the finite state graph of the concurrent system. The orientation of the method is to provide a practical verification method. The technical formulation of the Model checking is: Given structure M, state s, and TL formula f, does M, s | = f?. Clarke and Emerson [14] formulated the CTL (Computation Tree Logic) and proposed a CTL Model checking algorithm. They proposed that concurrent programs can be abstracted to finite state synchronization skeletons, suppressing behavior irrelevant to concurrency.

Model checking addresses finite systems but can be scaled up to a more complex system as a multi-agent system. Here, by complex we mean a system with a large number of independent interacting components, with non-linear aggregate activity, with concurrency between components and constant evolution. Model checking basic idea is to use algorithms executed by software tools to verify the correctness of the system. The user inputs a description of a model of the system, the possible behavior, and a description of the requirements specification, i.e. the desirable behavior, and leaves the verification up to the machine. If an error is recognized, the tool provides a counter-example showing under which circumstances the error can be generated. The counter-example consists of a scenario in which the model behaves in an undesired way. Thus the counter-example provides evidence that the model is faulty and needs to be revised. This allows the user to locate the error and to repair the model specification before continuing. If no errors are found, the user can refine its model description e.g. by taking more design decisions into account so that the model becomes more concrete and can restart the verification process.

## 2.6. Timed Automata

A timed automaton is a finite state automaton equipped with a finite set of real value clock variables called clocks, which are used to measure the elapse of time. Timed automata are used to model finite state real-time systems. A state of a timed automaton consists of the current location of the automaton plus the current values of all clock variables. Clocks can be initialized when the system makes a transition. Once initialized, they start incrementing their value implicitly. All clocks proceed at the same rate. The value of a clock thus denotes the amount of time that has been elapsed since it has been initialized. Conditions on the values of clocks are used as enabling conditions of transitions: only if the clock constraint is fulfilled, the transition is enabled, and can be taken; otherwise, the transition is blocked. Invariants on clocks are used to limit the amount of time that maybe spent in a location. Enabling conditions and invariants are constraints over clocks.

A timed automaton A is a tuple $(L, l_0, E, Label, C, clocks, guard, inv)$ [21] with

- $L$, a non-empty, finite set of locations with initial location $l_0 \in L$
- $E \subseteq L \times L$, a set of edges
- $Label: L \rightarrow 2^{AP}$, a function that assigns to each location $l \in L$ a set Label($l$) of atomic propositions
- $C$, a finite set of clocks
- $clocks: E \rightarrow 2^{C}$, a function that assigns to each edge $e \in E$ a set of clocks clocks(e)
  $clocks: E \rightarrow \psi(C)$, a function that assigns to each edge $e \in E$ a set of clocks clocks(e)
- $guard: E \rightarrow \psi(C)$, a function that labels each edge $e \in E$ with a clock constraint guard(e) over C, and
- $inv: L \rightarrow \psi(C)$, a function that assigns to each location an *invariant*.

# 3. UPPAAL

UPPAAL [20] is a toolkit for symbolic model checking of real-time systems developed at the University of Uppsala (Sweden) and Aalborg (Denmark). It provides model checking for verification of behavioral properties as well as simulation of timed automata. It also has some features to detect deadlocks. The model checking algorithms that are implemented in UPPAAL are based on sets of clock constraints, rather than on explicit sets of regions. By dealing with (disjoint) sets of clock constraints, a coarser partitioning of the infinite state space is obtained. A multi-agent system in UPPAAL is modeled as a network of timed automata, called processes. A clock variable evaluates to a real number and clocks progress synchronously. The fulfilled constraints for the clock values only enable state transitions but do not force them to be taken. A process is an instance of a parameterized template. A template can have local declared variables, functions, and labeled locations. State of the system is defined by locations of the automata, clocks, and variables values.

UPPAAL uses a query language (i.e. subset of TCTL) for defining requirements. The query language consists of state formulae and path formulae. State formulae describe individual states with regular expressions such as $x \geq 0$. State formulae can also be used to test whether a process is in a given location. Path formulae quantify over paths of the model and can be classified into reachability, safety, and liveness properties:

- Reachability properties are used to check whether a given state formula f can be satisfied by some reachable state. The syntax for writing this property is E <> f.
- Safety properties are used to verify that something bad will never happen. There are two path formulae for checking safety properties. A[] f expresses that a given state formula f should be true in all reachable states, and E[] f means that there should exist a path that is either infinite, or the last state has no outgoing transitions, called maximal path, such that f is always true.
- Liveness properties are used to verify that something eventually will hold, which is expressed as A <> f.

Processes communicate with each other through channels. Binary channels are declared as chan x. The sender x! can synchronize with the receiver x? through an edge. If there are multiple receivers x?, then a single receiver will be chosen non-deterministically. The sender x! will be blocked if there is no receiver. Broadcast channels are declared as broadcast chan x. The syntax for sender x! and receiver x? are the same as for binary channels. However, a broadcast channel sends a signal to all the receivers, and if there is no receiver, the sender will not be blocked. UPPAAL also supports arrays of channels. The syntax to declare them is chan x [N] or broadcast chan x [N], and sending and receiving signals are specified as x[id]! and x[id]?. Processes cannot pass data through signals. If a process wants to send data to another process then the sender has to put the data in a shared variable before sending a signal and the receiver will get the data from shared variable after receiving the signal.

# 4. CASE STUDY: A MULTI-AGENT TRANSPORT SYSTEM

In this section we present a case study of multi-agent system. It is a system composed of transporting agents. The objective is to specify our system using timed automata and then verify the correctness properties of safety and liveness. The mission is to transport stock from one storehouse to another. They move in their environment which in this case is static, i.e. topology of the system does not evolve at run time. We have specified each and every part of the system, i.e. agents along with the environment in the form of LTS.

There are three types of agents:

1) **Carrier agent:** It transports stock from one storehouse to another, it can be loaded or unloaded and can move both forward and backward direction. Each road section is marked by a sign number and the carrier agent can read this number.

2) **Loader/Un-loader agent:** It receives/delivers stock from the storehouse, it can detect if a carrier is waiting (for loading or unloading) by reading the presence sensor, it ensures that the carrier waiting to be loaded is loaded and the carrier waiting to be unloaded is unloaded.

3) **Store-manager agent:** It manages the stock count in the storehouse and it also transports the stock between storehouse and loader/un-loader.

## 4.1. UPPAAL Model

The template of the Carrier agent has eight locations: Safe, Appr Loaded Carrier, Parking Store House A, Start Loaded Carrier, Crossing, unload Carrier, Parking Store House B and Start Empty Carrier. The templates for Carrier agent and path have been modeled as shown in figure 1.

***The Carrier Agent Template:*** The initial location is Safe, which corresponds to a carrier agent has not appeared on crossing loaded yet. The location has no invariant, which means that a carrier agent may stay in this location for an unlimited amount of time. When a carrier agent is approaching, it synchronizes with the controller. This is done by the channel synchronization appr! on the transition to Appr Loaded Carrier. The controller has a corresponding appr?. The clock x is reset and the parameterized variable e is set to the identity of this carrier agent. This variable is used by the queue and the controller to know which carrier agent is allowed to continue or which carrier agent must be stopped and later restarted.

The location Appr Loaded Carrier has the invariant x ≤ 20, which means that the carrier agent must leave the location within 20 time units. The two outgoing transitions are guarded by the constraints x ≤ 10 and x ≥ 10, which corresponds to the two sections before the crossing: can be stopped and cannot be stopped. At exactly 10, both transitions are enabled, which allows us to take into account any race conditions if there is one. If the carrier agent can be stopped (x ≤ 10) then the transition to the location Parking Store House A is taken, otherwise the carrier agent goes to location Crossing. The transition to Parking Store House A is also guarded by the condition e == id and is synchronized with stop?

When the controller decides to stop a carrier agent, it decides which one (sets e) and synchronizes with stop!

The location Parking Store House A has no invariant: a carrier agent may be stopped for an unlimited amount of time. It waits for the synchronization go?. The guard e == id ensures that the right carrier agent is restarted.

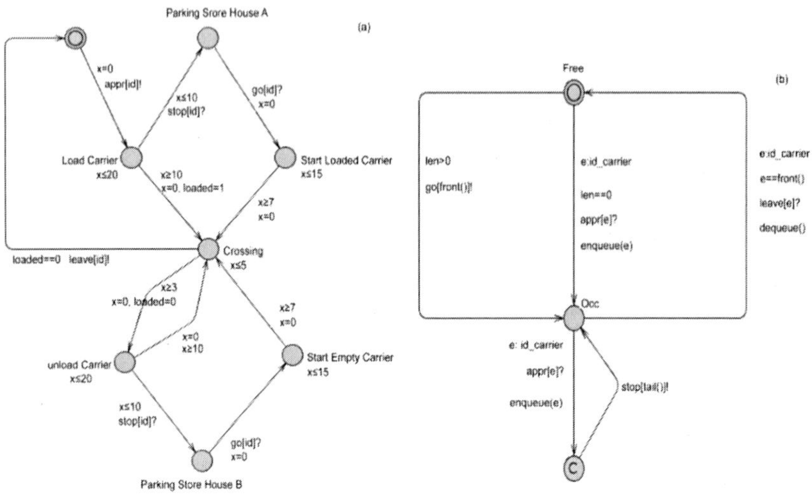


**Figure 1**. The template for the (a) Carrier agent; and (b) Carrier path.

We can assume that a carrier agent may receive a go? synchronization even when it is not stopped completely, which will give a non-deterministic restarting time. The location Start Loaded Carrier has the invariant $x \leq 15$ and its outgoing transition has the constraint $x \geq 7$.

This means that a carrier agent is restarted and reaches the crossing section between 7 and 15 time units non-deterministically. The location Crossing is similar to Start Loaded Carrier in the sense that it is left between 3 and 5 time units after entering it.

***The Template of the Path:*** The path controller synchronizes with the Carrier agent. Some of its locations do not have names. Typically, they are committed locations (marked with a c). The controller starts in the Free location (i.e., the path is free), where it tests the queue to see if it is empty or not. If the queue is empty then the controller waits for approaching carrier agent with the appr? synchronization. When a carrier agent is approaching, it is added to the queue with the add! synchronization. If the queue is not empty, then the first carrier agent on the queue is restarted with the go! synchronization.

In the Occ location, the controller essentially waits for the running carrier agent to leave the path (leave?). If other carrier agent is approaching (appr?), they are stopped (stop!) and added to the queue (add!). When a carrier agent leaves the path, the controller removes it from the queue.

**Table 1**. UPPAALspecifications (liveness and safety properties).

| Declarations | UPPAAL Specifications |
|---|---|
| System declarations | //Template instantiations<br>system Carrier_Agent,Path; |
| Global declarations | //Global declarations<br>const int N = 2; //number of carrier agents<br>typedef int [0,N-1] id_carrier;<br>chanappr[N], stop[N], leave[N];<br>urgent chan go[N]; |
| Carrier agent declarations | id_carrier list[N+1];<br>int[0,N] len;<br>// Postcondition: Put an element at the end of the queue<br>void enqueue(id_carrier element)<br>{list[len++] = element;}<br>// Postcondition: Removes the front element of the queue<br>void dequeue()<br>{<br>int i = 0;<br>len -= 1;<br>while (i < len)<br>{list[i] = list[i + 1];<br>i++;}<br>list[i] = 0;<br>}<br>// Postcondition: Returns the front element of the queue<br>id_carrier front()<br>{return list[0];}<br>// Postcondition: Returns the last element of the queue<br>id_carrier tail()<br>{return list[len - 1];} |
| Formal verification using UPPAAL verifier | //Collion Detection on Crossing Area<br>E<> Carrier_Agent(1).Crossing imply not Carrier_Agent(0).Crossing<br>//Reachability<br>E<> Carrier_Agent(1).Crossing<br>//Check Deadlock For System<br>A[] not deadlock |

Table 1 shows the UPPAAL code snippets for system declarations, global declarations, and carrier agent declarations. There is also the UPPAAL verification code which highlights the safety properties.

# 5. CONCLUSIONS AND FUTURE WORK

A well-defined, precise, timed automaton based model of a multi-agent transport system is proposed. The correctness properties of safety and liveness of this proposed model are verified. The multi-agent system is modeled as a network of timed automata. A clock variable evaluates to a real number and clocks progress synchronously.

Our future work is the proposition, design and implementation of transformation rules for the translation of the current timed automata based formal model into a working system. This working system would have a formal foundation as it would be based on timed automata model.

## ACKNOWLEDGEMENTS

We are grateful to The Worthy Vice Chancellor, The Islamia University of Bahawalpur for motivation and en couragement. This work has been possible due to the support of The Department of Computer Science & IT, The Islamia University of Bahawalpur, Pakistan.

## REFERENCES

1. George, V. and Vaughn, R. (2003) Application of Lightweight Formal Methods in Requirement Engineering. Crosstalk: The Journal of Defense Software Engineering, 16, 30.

2. Wooldridge, M. (2002) An Introduction to MultiAgent Systems. John Wiley and Sons, Chichester.

3. Wooldridge, M., Jennings, N.R. and Kinny, D. (2000) The Gaia Methodology for Agent-Oriented Analysis and Design. Autonomous Agents and Multi-Agent Systems, 3, 285-312.

4. Dogac, A. and Cingil, I. (2004) Agent Technology. In: B2B E-Commerce Technology: Frameworks, Standards and Emerging Issues, Addison-Wesley, Boston, 103-150.

5. Wooldridge, M. and Jennings, N.R. (1995) Intelligent Agents: Theory and Practice. The Knowledge Engineering Review, 10, 115-152.

6. Weyns, D., Omicini, A. and Odell, J. (2007) Environment as a First-Class Abstraction in Multi-Agent Systems. International Journal of Autonomous Agents and Multi-Agent Systems, 14, 5-30.

7. Russell, S. and Norvig, P. (2002) Artificial Intelligence: A Modern Approach. 2nd Edition, Prentice Hall, Upper Saddle River.

8. Demazeau, Y. (2003) Multi-Agent Systems Methodology. Franco-Mexican School on Cooperative and Distributed Systems (LAFMI).

9. Ferber, J. (1999) An Introduction to Distributed Artificial Intelligence. Addison-Wesley, Boston.

10. Jennings, N.R., Sycara, K. and Wooldridge, M. (1998) A Roadmap of Agent Research and Development. International Autonomous Agents and Multi-Agent Systems, 1, 7-38.

11. Ghezzi, C., Jazayeri, M. and Mandrioli, D. (2003) Fundamentals of Software Engineering. 2nd Edition, Prentice Hall, Upper Saddle River.

12. Magee, J. and Kramer, J. (2006) Concurrency: State Models and Java Programs. 2nd Edition, John Wiley and Sons, Hoboken.

13. Giannakopoulou, D., Magee, J. and Kramer, J. (1999) Fairness and Priority in Progress Property Analysis. Technical report, Department of Computing, Imperial College of Science, Technology and Medicine, London.

14. Clarke, E.M. and Emerson, E.A. (1981) Design and Synthesis of Synchronization Skeletons Using Branching Time Temporal Logic. In: Kozen, D., Ed., Logics of Programs, Volume 131, Springer-Verlag, New York, 52-71.

15. Quielle, J.P. and Sifakis, J. (1982) Specification and Verification of Concurrent Systems in CESAR. Proceedings of the 5th International Symposium on Programming, Turin, 6-8 April 1982, 337-351.

16. Clarke, E., Grumberg, O. and Peled, D. (1999) Model Checking. MIT Press, Cambridge.

17. Clarke, E.M., Emerson, E.A. and Sistla, A.P. (1986) Automatic Verification of Finite State Concurrent Systems Using Temporal Logic Specifications. ACM Transactions on Programming Languages and Systems, 8, 244-263.

18. Clarke, E.M., Grumberg, O. and Long, D.E. (1994) Model Checking and Abstraction. ACM Transactions on Programming Languages and Systems, 16, 1512-1542.

19. Clarke, E.M., Grumberg, O., Jha, S., Lu, Y. and Veith, H. (2003) Counter Example-Guided Abstraction Refinement for Symbolic Model Checking. Journal of the ACM, 50, 752-794.

20. Larsen, K.G., Pettersson, P. and Yi, W. (1997) UPPAAL in a Nutshell. International Journal on Software Tools for Technology Transfer, 1, 134-152.

21. Katoen, J.P. (1999) Concepts, Algorithms, and Tools for Model Checking. A Lecture Notes of the Course Mechanized Validation of Parallel Systems. For 1998/99 at Friedrich-Alexander Universitat, Erlangen-Nurnberg, 195.

# INDEX